DOING SCIENCE

The Reality Club

John Brockman
Editor

PRENTICE
HALL
PRESS

New York   London   Toronto   Sydney   Tokyo   Singapore

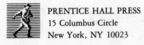 PRENTICE HALL PRESS
15 Columbus Circle
New York, NY 10023

PRENTICE HALL PRESS and colophons are registered trademarks
of Simon & Schuster, Inc.

Library of Congress Cataloging-in-Publication Data

Doing science / John Brockman, editor.
        p. cm.—(The Reality Club ; 2)
        A collection of original articles written by members of the Reality Club.
        Includes index.
        ISBN 0-13-795097-7
        1. Science. 2. Technology. I. Brockman, John, 1941–   ﹅
II. Reality Club. III. Series: Reality Club (Series) ; 2.
Q158.5.D65   1991
500—dc20   90-37153
                                                              CIP

Manufactured in the United States of America

10 9 8 7 6 5 4 3 2 1

First Edition

For information about our audio products, write us at:
Newbridge Book Clubs, 3000 Cindel Drive, Delran, NJ 08370

Acknowledgments

Lynn Margulis acknowledges the aid of Professor R. Guerrero, Dr. G. R. Fleishaker, Sam Beshers, Tom Lang, and Dorion Sagan in preparation of this manuscript. Support for her work comes from the Richard Lounsbery Foundation, New York City, the Evolution Fund of Boston University, the Botany Department of the University of Massachusetts at Amherst, and the NASA Life Sciences office (grant NGR 004-025). She thanks John T. Kearney and Rene Fester for manuscript preparation.

Financial support for the work of Francisco Varela and Antonio Coutinho was provided by the Fondation de France and the Prince Trust Fund.

The preparation of Robert Sternberg's piece was supported by a grant from the Guggenheim Foundation, a contract from the Office of Naval Research, and a contract from the Army Research Institute.

Contents

Designing Perpetual Novelty: Selected Notes from the Second Artificial Life Conference

KEVIN KELLY

Brooks has plans to invade the moon with a fleet of shoebox-size robots that can be launched from throwaway rockets. It's the ant strategy. Send an army of dispensable limited agents coordinated on a task, and set them loose. Some will die, most will work, something will get done. In the time it takes to argue about one big sucker, Brooks can have his invasion built and delivered. The motto: Fast, Cheap, and Out of Control.

The Second Artificial Life Conference was held in Santa Fe from February 5 to 9, 1990; it was a follow-up to the first Artificial Life Conference held two years before. Both conferences gathered an eclectic cross-disciplinary group of researchers including biologists, computer scientists, ecologists, astronomers, mathematicians, computer graphics experts, geneticists, roboticists, and avant-garde science groupies, of which I am probably one. The conference lasted a week. This report follows a rough chronological order of the week as it unfolded, with the exception of the few talks that I missed.

═══

David Campbell made the welcoming speech. He suggested the endeavor of the collective group was to make a "synthetic biology." He described artificial life (A-life) as a "natural" effect, a further evolution, the introduction of new diversity with new models of life.

═══

Chris Langton, the organizer of the first A-Life Conference and co-organizer of this one, gave the opening talk. He was the only one of the week to dare offer a definition of A-life. His try: "The attempt to abstract the logical form of life in different material forms." His thesis is that life is a process, or relationship, or logic, or complexity that is not bound to a specific material manifestation. He feels that even a mild A-life success enables us to study life in such a way that we can disassemble natural systems as we cannot do in nature, either practically or morally. The main task he sees before us is to discern how the macroscopic emerges dynamically from the microscopic, how a higher level of organization emerges from a lower one.

═══

Tom Toffoli, master cellular automata (CA) pioneer. He likes to tell
parables in his lilting Italian accent. His stories seemed to be Italian
folktales rather than the cybernetic koans they were. He told of
being on vacation somewhere once, idly watching ants in his room
try to haul away a dead spider they had captured. The crack in the
wall the ants had come through was too small for the spider's body,
so they dragged the monster up to the window sill. But the crack in
the window sill was also too narrow, so they dragged the huge spider
down to the crack in the wall again. Naturally it was no bigger the
second time around, so they hauled the spider up to the window
again, and back down again, and so on. Toffoli suddenly realized
that the movements of the group of ants was a periodic oscillation.
Their group behavior operated on a cycle quite different from any of
their individual behaviors, or even the behavior of the colony as a
whole. He hinted that what artificial life will have, and what CA
must acquire, is multiple scales of velocity.

His was the first talk to point to a phenomenon brilliantly explained by the ecologist R. V. O'Niell in his new book, *The Hierarchial Concept of Ecosystems*. Ecology, and presumably A-life, is
governed by hierarchies of cycles, some fast, some slow. What makes
an ecology interesting and robust is the presence of many different
rates, and the unpredictable merging and cancellation in the inter-
sections of these waves. This variety of rates is as important as the
variety of genes. Ideally, he says, there must be at least 10^4 difference in scale between levels of organization in order to have the
richness humans find "interesting."

The other challenging thing I remember him saying is, "We can
usually expect long periods of steady state in most systems. A typical 'interesting' CA will go from the mostly boring to a sharp peak
of interesting activity, then back to the boring. If you look on a
large scale, even life as we know it would seem to be a steady state.
So what is wrong with steady state?"

More detail than we can use is one goal of A-life. Humans are too
stingy in granting the generous levels of complexity an artificial life
needs. Toffoli quoted Norm Margulis, another CA experimenter, as
saying the remarkable thing about nature is the vast amount of

wasted details it produces—all those intricately executed leaves falling to decay on the ground.

His closing remarks suggested that although the mission of A-life can be described as the creation of a "synthetic biology," he prefers to view his work as the attempt to create a set of plastic initial conditions from which new worlds can be built, a "domesticated physics." Toffoli is working on a "programmable matter machine," a new architecture of supercomputing a thousand times more complex than the most complex existing supercomputer. His, the CAM-8, will allow the detail and multiple levels of activity he wants for A-life.

▭▭

Chris Langton spoke again, this time reporting on his research. His talk set the theme for much that later happened during the week. He began with a short history of the first conference, explaining that it was self-organizing. The usual way to organize a conference, particularly for its inaugural session, is to draw up a list of key researchers and do whatever is necessary to persuade them to come. For the A-Life Conferences, Chris merely came up with the title Artificial Life Conference and broadcast it into the scientific community to see who it would reel in. Those who came were the key people and their subjects the key fields. A-Life II worked the same way. There were fifty participants and four press at the first, and about two hundred fifty participants and twenty-five press at the second.

Langton works with cellular automata, huge grids of square cells imagined in computer space. Cells can be either dead (off) or alive (on) depending on complicated rules about their neighboring cells. Since a cell's neighbors determine its state, and its state determines the next generation of its neighbors, there is an ecological aspect to these worlds. This reciprocal relationship causes the cells to "behave" in a manner similar to simple living cell cultures of bacteria. Colonies form and disappear, some last many, many generations, some die out quickly, some go through endless repetitive cycles. The biological model of CA has been obvious since the first A-Life

Conference. What Langton brought to this one was a startling physical model.

By running thousands of CA under every possible rule he could
find within certain parameters, and then measuring one particular
aspect of their behavior—whether or not they had complex structures without being repetitious—he was able to map out the territory of this self-organizing behavior. He then graphed the results,
which formed a narrow Mount Fuji. Starting at one side, there were
the fixed and steady-state worlds, which were homogeneous and
stagnant. Increasing the x lead to more heterogeneity, but also more
periodic and repetitious patterns. Further up came a sharp peak of
complex, local structures, which dropped off sharply into a nonperiodic chaos. The shape of that graph nicely matches a common
phase-change graph for most elementary materials. Iron, or oxygen,
will vary from solid to liquid to gas depending upon temperature
and pressure. What Langton suggests is that CA, perhaps A-life as
well, can be thought of as existing within the continuum of a phase
transition, as a sharp, thin line between periodic or static routines
(solids) and nonperiodic chaos (liquids). This gave rise to a common
call during the week for "life being lived on the edge of chaos."

A secondary remark that Chris made in passing was that "states
of matter might be states of behavior." And going along with
Toffoli, he suggested that A-life could be thought of as finding the
universal class of artificial matter, that is, extending the idea of what
solids and liquids and gases are.

Another point about his phase-transition graphs, which I don't
fully understand, is that the two sides of the Mount Fuji graph can
be thought of as competing with each other when learning is introduced into the rules. One could say alternatively that learning forces
optimization, pushing it toward the peak, which is the optimum
domain of information use. So when learning rules are introduced
they drive against both chaos and stagnation toward the thin summit of the phase transition. Langton put it this way, using the newly
minted vocabulary of chaos theory: "Evolution is learning how to
avoid the strange attractors of phases."

What interested me was how often in the conference participants
made this equation: evolution = learning.

Using these learning CA, Langton foresees achieving a deeper kind of learning. He's trying to construct initial local rules for CA so that by learning optimization "they will converge on better local rules that will compute an answer," rather than having to hard-wire the mechanism for getting answers. This is also the goal shared by the neural-net field.

Steen Rasmussen, a Dutch scientist, was so taken by the demonstrations of core war games at the first A-Life Conference that he has since used the massively parallel Connection Machine to explore core wars big time. A core war is the competition between two simple codes in an empty computer's core. Each side is a elementary program whose single intent is to replicate over the entire memory space, including whatever space the other bug and its replications are taking up. In ecological terms, two critters with different genes are competing for the same limited resources (computer memory) and can also consume each other and their progeny. The fittest takes over the entire world. In one way a core war could be thought of as a deliberate battle between computer worms in a controlled space within one computer.

The Connection Machine offers a truly vast universe to play core wars in. There are $10^{33,000}$ possible states in Rasmussen's Virtual Evolution in a Nondeterministic Universe Simulation or VENUS, compared to only 10^{80} particles in the real universe. By letting the machine create worms and running them against each other again and again in endless variation he finds that stability and complexity tend to cluster around certain rules. When mapped, these rules form basins of stability separated by ridges of chaos, watersheds of complexity. He concludes that "stability breeds diversity," which is the inverse of "diversity breeds stability," the traditional ecological wisdom.

Stuart Kaufman is one of the "founders" of A-life, although I may be the first to call him that. There are few theoretical biologists in the world; he is one of them. (There are also few recipients of the

MacArthur Award, the "genius" prize for independent work, and he is one of those, too.) Kaufman's search is best summed up by the title of the book he is just finishing, *Origins of Order: Self-Organization and Selection in Evolution,* to be published this year by Oxford University Press. To him, A-life is merely the application of theoretical biology. It is the lab work for theories of evolution.

He says Darwin was right as far as he goes, but that Darwin had no idea that complex systems of all kinds exhibit self-organizing properties, so that he left the details of how selection works unspecified. The lack of understanding of how selection works is the biggest gap in biological science currently. Kaufman sees life as an *"expected,* emergent property of self-organized systems." Coming from someone else this might be plain mysticism, but Kaufman is primarily a mathematician and has numbers increasingly backed up by simulations created by others as evidence for his ideas. He works a lot with genetic algorithms because, as he says, "the genome is a parallel processing network, with 100,000 genes turning themselves on and off." Viewing DNA as a computer, as he does, means that inverting the metaphor and viewing a computer as DNA is not such a crazy notion.

He finished his very technical talk with a discussion of something he calls the Coevolution Avalanche, by which I believe he means the self-amplifying way in which coevolution proceeds to push organisms to optimize a niche. As in the arms race, the rate of coevolution increases until it becomes an avalanche, sweeping the coevolving community into greater coevolution. This idea emerged from his work with genetic algorithms and fitness landscapes—computer models of adaptation.

<hr/>

At lunch break there was a press conference. Chris Langton made some interesting statements. In response to the question, "What is the use of studying (creating/messing around with) artificial life?" he listed four reasons for such research:

- A-life gives a picture of nature as a whole. (And other things as a whole, I must add.)

- We need to study A-life because it is inevitably going to be with us. Look at computer viruses as an example.
- A-life is a better way to engineer complex software; if you can't build it, you can evolve it.
- A-life is a means to study biological life, which resists understanding as a historical case, and resists (practically and morally) altering a few parameters in a coldly scientific way.

When asked, "How did A-life research start?" Langton replied, "Apple Computer allowed the A-life field to be possible. A lot of the early work could be done on a small PC. The thing that impresses me even now is that nature has more computing power than we do." (If that last remark isn't representative of a paradigm shift, I don't know what is.)

"What are some of the questions you want to answer with A-life?" was the next question. Someone answered: "What governs systems that improve? What are the origins of intention and purpose?" One of the researchers added, "The technology of life will be very valuable. That's what A-life is—the technology of life."

Vladimir Kuz'min's strong Russian accent made his provocative talk difficult to follow. He is concerned with "breaking symmetries," and describes the history of the universe from the Big Bang onward as a series of symmetries that sprang into new asymmetrical forms. It's fairly well established that most biochemical molecules have "handedness," that is, their mirror reflections are not clones. Kuz'min claims that "handedness" (he also calls it "chiral purity") is a prerequisite for self-replication. That's an enticing, though vague, idea.

Ron Fox talked about the genesis of early life. His research investigates the energy cycles in proto-cells. He is trying to determine "what came first, the polymer or the energy." Both the proteins and the energy paths they need are sophisticated and unlikely to exist one without the other, even in the most simple forms of life. So how does it happen? Like all whole systems, nothing comes first. The only way to make a sophisticated whole system is to grow it from a simpler one. Together with Lynn Margulis's idea that the origins of cells lie in the symbiosis of three or four smaller proto-cell parts, I think that coevolution can be applied to the insides of an organism. Maybe it should be called cogenesis.

Doyne Farmer has passionate opinions and emotions about artificial life. His theme was "autocatalytic" action. By autocatalytic I think he means positive feedback, or self-amplification. He has discovered several effects, which make a feedback loop.

- Autocatalytic pathways allow "storage space" for complexity and diversity.
- Diversity breeds robustness.
- Robust autocatalytic pathways are self-healing.
- Diversity in a system increases autocatalytic processes.

I don't know whether this is tautological or cybernetic.

Peter Schuster is modeling the way in which a genotype (the program) will create a phenotype (the carrier). In particular he is using a simplified DNA that will generate, on a computer, a protein folding in two dimensions. One slide showed how radically different in form a single difference in their codes will make two proteins, almost the difference between a circle and a cross. In a simulation of evolution he applied selection pressures to the phenotypes created by small mutations in the genotype to see if this small world would learn to adapt. He reported that increasing the error (mutation) rate would keep increasing evolvability until a critical "error threshold" is reached, at which point further evolvability becomes difficult. This suggests that mutation rates are optimized. His threshold was, I think, 1 percent, or in that range. There are a few other A-life experiments (including Richard Dawkins's Biomorph world in A-Life I) that have come up with mutation rates. Somebody should make a comparison table of them to see if they converge.

Following up on some of Ron Fox's ideas concerning the origins of life, Gerald Joyce has been investigating primordial soups. The central dogma of molecular biology is that DNA produces RNA, which builds the proteins that compose the phenotype, which in turn transfers selection pressures to the mutations, thereby altering the DNA which produces RNA, and so on. Joyce's work suggests that you can eliminate both DNA and the protein from that loop, giving us an autonomous RNA-based world. This is important news in the investigation of the origins of biological life (both prehistoric and artificial) because it provides a scenario of how such a complicated DNA genome/phenome pathway could have arisen through a solitary intermediary. In fact, Joyce has made this RNA world work in the lab. He found particular RNA sequences that would serve both as code and vehicle so that the new (old) loop went as follows: RNA, which composes the phenotype, which transfers selection pressures to the mutations, thus altering the RNA, which composes the phenotype, and so on. Initially he designed each step in separate containers, but he has recently gotten the whole RNA cycle to work in one container, one soup.

David Jefferson, from UCLA, showed the first of many ant worlds being premiered at the conference. Later in the week, another ant-world enthusiast began quietly handing out tiny rubber ants ordered from the zany Archie McFee mail-order catalog of rubber chickens and other novelties. Each day more and more conference participants were walking around wearing name tags crawling with carpenter ants under the plastic slip. By the end of the week, the ants had become the mascot of the conference.

Ants were independently selected by a number of researchers as ideal A-life models because they are such a handy and nonthreatening example (unlike cockroaches) of simplicity generating complexity. To cyberneticists, ants are cute.

Jefferson created a computer-modeled world called Genesys to try out artificial evolution. His ant creatures are neural-net animals;

they have simple algorithms that tell them how to move or turn. The only resource they consume is decision (or computing) power. Their only goal in life is to find their way through a very complicated virtual maze. Since they have limited decision resources, they not only can't afford to make many wrong moves but they can't spend too much time thinking about which move to make either. In other words, if they can figure out the few algorithms (rules of thumb) that get them through the maze without much thinking or error, they will succeed.

Jefferson has introduced a way for the ant's algorithms to mutate, generating new algorithms, and allowed the maze to be a selection pressure. Those randomly generated ants that score highest in traveling through the maze are kept to be re-released into it, and so on, for many generations. The brightest randomly generated ant could only figure out how to get through two-thirds of the maze before running out of thinking power. But after a hundred generations of evolution and sifting, a highly evolved ant could whip through the maze with a perfect score. Here it was not the humans but the artificial ants who developed the perfect rules of thumb.

This experiment was done with a fixed environment—the maze was the same the whole time. What happens when this army of highly evolved ants is put into a new maze? Jefferson and associates found they didn't fare well, particularly at first. After a while, the ants did learn how to go through better. But the surprising thing was that when a fresh set of randomly generated know-nothing ants was run instead of the highly evolved ants, the untrained ants reached a perfect score sooner. If I remember correctly, the specialized ants were never able to learn the new maze perfectly, at least in the finite number of generations run. It's the classic lesson of the dangers of overspecialization.

In ecology this is called the problem of local optimization, and it comes up often in A-life. Imagine we were to generate a "landscape" for the adaptive abilities of an organism, played by you. The more highly adaptive you, the organism, are, the higher the elevation. This landscape will look rugged; there will be many mountains and hills because your adaptive potential depends upon the outside environment. At any one time you are somewhere on this "rugged adaptive landscape" trying to climb to higher optimization. Sometimes there will be a peak that will stand high around the surrounding area, but which is much lower than a really high peak someplace else. In order to get over to that other higher peak you actually have to descend to a lower unoptimized state. It may be that you need to unravel so much that you cannot do so, being such an optimal organism. So you get stuck on a local high. If the environment shifts, you're doomed.

That's what happened to the specialized ants. They got caught on the local optimization of the first maze. How does an organism acquire generally adaptive behavior? This is the one of several Holy Grail questions in A-life.

Jefferson listed other questions he hopes that A-life can answer:

- What determines rates of mutation and recombination?
- How does cooperation evolve?
- How does sex evolve?
- How does speciation happen?
- Is there anything to Lamarck's thesis that acquired traits are inherited?

He said that what he had learned so far from the ant worlds in Genesys is that "Evolution is massively parallel learning."

━━━

Early in his talk, Danny Hillis, the inventor of the Connection Machine, gave the formula for inventing sex. He titled it "How to make sexual evolution":

- Make a random start.
- Score fitness.
- Select survivors.
- Choose mates.
- Combine genes.
- Mutate next time around.

This was the first time A-life sex was introduced. In the context of the week, the participants found it pretty exciting.

Hillis's formula came from his investigations on the Red Queen System, a ecological model based on his own Connection Machine. This supercomputer uses 64,000 processes in parallel (versus the one or two in most computers) to simulate an interacting world. Each organism in the Red Queen System is modeled by one dedicated processor. Therefore each organism can perform its own independent interaction with other similar organisms. The combined ecology of 64,000 reciprocating organisms is what makes the Red Queen System.

It is an evolving sexual world. The organisms are "sorting networks," virtual beings whose task in life is to perform calculations. Their fitness is scored on how well they solve numerical problems. Those performing best survive to pass their rules onto the next generation. Introducing sex speeds up the process of attaining fitness. Yet Hillis discovered something that speeds up the fitness process even more: parasites.

By introducing a second kind of organism to his small worlds, Danny found that the system exhibited many more interesting levels of organization and behavior. This new organism would live off the bounty of thriving prime organisms, weakening them, but not killing them off. To thrive, prime organisms now had two tasks, to solve calculations better, and to become less attractive to parasite organisms. However, because parasites were also evolving in the system, finding new opportunities to rely on prime organisms, this parasitism became a dynamic selection pressure keeping the whole system in flux. It is from this constant race to stay in place—the lament of the Red Queen in *Alice's Adventures in Wonderland* that it takes all the running she can do to stay in the same spot—that the Red Queen System gets its name.

Hillis sees A-life as offering a new interpretation of biology. He says the reigning dogma is that the "natural order" specifies certain roles for organisms in nature. Oak trees should be protected because they do this or that in a forest, and oak forests should be protected because they do this or that for a certain area. But you can't separate an oak tree from the forest, or a forest from a biome. He says ecologists and perhaps environmentalists are beginning to understand that "oak tree," "oak forest" are not only fluid and continually being reinvented but are almost a phantom as separate entities. Like Richard Dawkins, of *The Selfish Gene* and *Extended Phenotype* fame, Hillis says "ideas of independent genes are illusory." An "oak tree" includes all the parasites that keep it going evolutionarily, and vice versa. It's a perspective of ever-widening circles of symbiosis.

To my mind one of the most remarkable findings from the Red Queen System is Danny's graph of his organism-with-parasite's rise in fitness over time. Over a thousand generations, their fitness mildly increases, then zooms up precipitously, then levels off for a while, then zooms up rapidly again, then levels off. If we understand fitness as adaptation, this graph appears to be a spitting image of the recent theories of punctuated equilibrium in evolution promoted by Stephen Jay Gould and others. By and large, they argue, evolution proceeds at a pace near equilibrium, which is occasionally broken by intense periods of readjustment and rapid change. Hillis's evolutions showed the same, with longer periods of equilibrium punctuated by shorter, quick spurts of increased adaptation toward fitness.

Theoretical biologists drool over the prospect of messing around with synthetic evolution tools like this, but Hillis has real-world applications in mind, too—flying airplanes and such. Hillis sees evolving ecologies like this being able to design things humans may not have the patience to solve. "We want these systems to solve a problem we don't know how to solve, but merely know how to state." The idea is to grow solutions. Set up a system that will evolve programs that will solve the problem you have at hand. Hillis: "Rather than spending uncountable hours designing code, doing error checking, and so on, we like to spend more time making better parasites." Better parasites mean faster convergence of the prime rule-making organisms to the fitness ideal—an error-free, robust software program. "I would rather fly on a plane that was based on software built by a program like this, than on software that I wrote myself, because it would be built in an environment with thousands of adversaries who specialize in trying to find what's wrong with it, so that whatever survives has been tested ruthlessly."

Alvy Ray Smith presented a paper he wrote before cofounding Pixar, the acme computer graphic company. It was a highly mathematical proof of the smallest "simple nontrivial self-reproducing machines." What I remember was his wistful memoir of how he became involved with computer graphics. He was persuaded by the work of biologists such as D'Arcy Thompson that growth was a computational design. Here Smith showed famous pictures of various species of fish plotted on a distorted grid to show similarities. In particular he was sure that the growth of an embryo is a computational program. He set out to write some tools to show this, and along the way invented major computer-graphic techniques such as ray-tracing, but never got back to the embryos. When he was at Lucasfilm he did do some graphic work with L-systems, one factor in computational growth, including the slide that began his talk, of a clump of wild grasses in White Sands National Monument. He ended with a call to the current generation of A-lifers to tackle the challenging problem of morphogenesis.

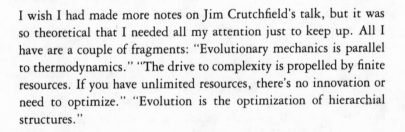

I wish I had made more notes on Jim Crutchfield's talk, but it was so theoretical that I needed all my attention just to keep up. All I have are a couple of fragments: "Evolutionary mechanics is parallel to thermodynamics." "The drive to complexity is propelled by finite resources. If you have unlimited resources, there's no innovation or need to optimize." "Evolution is the optimization of hierarchial structures."

Eugene Spafford, computer-security expert, gave a rundown on the current state of inadvertent A-life, the worms, viruses, bacteria, and other creatures on the loose in computer networks. He started with some definitions:

- *Worms* propagate, or move over networks, and may perform other actions besides replication.
- *Bacteria* (or rabbits) merely replicate in a known host.

- *Viruses* insert themselves into existing programs, but cannot be run on their own, and spread by replicating.

Spafford now has records of more than 115 versions or species of computer virus. Some viruses have become quite sophisticated. At least one pair of viruses (NVIR-A and NVIR-B on the Mac) have been known to "mate" by over-writing code to produce a strain more virulent than either. There are also cases of viruses able to detect the signature of other viruses present on the system. These aggressive viruses remove the first virus, and then insert themselves. Removal of the first virus lessens the chance of the second's being detected.

Other tricks abound. In response to more wary computer operators who try to wipe them out, some viruses will fake a reboot by dwelling in the memory. Upon discovering the presence of these memory-resident viruses, the operator will attempt to kill them or clean the system by turning the computer off. Click. Blank screen. Operator turns computer back on. Fresh screen, fresh memory, no more virus.

Wrong.

The virus, anticipating these moves, has control of the system and merely mimics the effects of turning off the memory, without letting it really happen. While pretending to be erased, the memory is still alive and holding the virus. Sort of like playing 'possum.

Russell Brand was probably the most entertaining speaker of the week. His was the only talk that was "interactive," demanding participation from the audience. He described a puzzling case he and other computer-security experts had encountered, and asked for guesses as to the agent. His thesis was that it is impossible to tell the difference between a human and a computer virus. By the end of this talk, almost everyone agreed with him.

The case involved abnormal behavior of a computer system. The system administrators noticed unofficial log-in attempts; messages left on the system, with more and more over a period of time that were identical including a spelling mistake; then more machines infected by messages without the spelling mistake; then an increasing number of messages left on many sites at exactly the same moment, and so on. A detective story. Who dunnit? A virus that misspells (on purpose to mislead?), or a human who can be in more than one place at once, or, a conspiracy? The point Brand wanted to make was that they had no idea whether they were dealing with bugs or people.

The ending of the whodunnit is that a meme—an idea that passes itself around and infects people—was responsible. In this instance it was a message written on a piece of paper that was passed around by students. This accounts for both the misspellings and the simultaneous spreading entries.

Russell Brand had some serious points, too. He made a very convincing case that there is nothing anyone would want to do for which a virus is the best means. He took challenges from the floor (cheap way to distribute software, as a means of hi-tech warfare, etc.) and gave clever and witty replies to all of them. During this exchange, Danny Hillis asked if the proliferation of UNIX standard machines might be causing more viruses. "Definitely," Brand replied. "In fact some people are deliberately staying with antiquated nonstandard versions of UNIX in order to remain immune from these common infections."

The problem we have with computer viruses, Danny said, stems from the fact that all our operating systems are identical. The very thing that has made computing easy for the user, a standard system, has made it easy for viruses. There is a continuing move to standardization among machines connected by networks. So as long as formats like UNIX become a universal standard, we'll have awful problems with viruses no matter how many vaccines and quarantines we come up with. What we want in networked computing is a diversity of operating standards. We want each computer to be a slight variant of the standard, maybe one that is slowly evolving. It will still have many holes that can be exploited by viruses and so forth; it won't be any more immune to infections, but it will hardly be worth the time to try to infect just one machine.

Danny made me realize that we have monocropping in computers. The idea of having a computer with an adapting operating system, one that is slightly different from all others, is both poetic and frightening. This way the computer becomes more like a pet with individual character, and (this is the scary part) with unique likes and dislikes. Just when computers were becoming manageable because they were predictable, we find that ultimately that very predictability will be their undoing. To quote William Safire (speaking of the dangers of peaceful stability in politics), "Hail to Unpredictability!"

There was a panel discussion about the implications of viruses as an A-life form. Harold Thimbleby from Scotland outlined a serious proposal to use a wormlike mechanism to distribute software updates. LiveWare he calls it. The engineered selective worm is broadcast out into the world; when it finds a receptive host, the worm infects it with an updated version of information or software. The key here is that the worm is selective, only entering those systems that have purposefully allowed it, and passing over those who do not have the needed welcome signal. It is, in effect, a self-distributing system, since the sender has no need to know to whom or where to send his information. In rebuttal, Russell Brand pointed out the dangers of such plans and continued to claim that "for any goal viruses are the wrong mechanism." Eugene Spafford also noted that so far "no computer virus has gone extinct." Panel member biologist Hyman Harthman dryly noted that before we dismiss them out of hand we should keep in mind that viruses and related organisms form the bulk of living matter on earth. Furthermore, he suggested, there has been a recent "theory of speciation by infection," which says that interspecies viral infections are what spurs the movement toward distinct germ lines, and that the crossover code from viral infections helps speed evolution. If I understood him correctly, he also said that researchers have noticed that interspecies viruses moving in the germ line are a steadily increasing phenomenon in living organisms on earth right now.

That was the closest anybody would come to endorsing viruses as a legitimate research area for A-life. The same amazing thing happened at the Hackers' Conferences. Not even there would anyone publicly defend experimenting with viruses. Privately, every hacker I talked with would say that viruses were fascinating conceptually, that they were important, if not inevitable, but that they were "wrong." Here too, scientists would confess privately to me their fascination with virus code and their desire to try something. Occasionally they would describe a design for a virus that they had in mind and would like to check out, but add, "Of course, I wouldn't do that!" Publicly, they sat mute while the virus bashers railed. I was seeing a twenty-first century taboo arise.

Steven Levy, author of *Hackers,* who was sitting with me, was getting upset. "I don't understand. Here we are at a conference on the making of A-life, and the closest example that we have of that, computer viruses, nobody will even stand up for. If they can't deal with it at this stage, how are they going to deal with full-grown artificial life?" I felt equally disgruntled. Biologically, viruses are more important to what happens on earth than dogs or cats. I wish Lynn Margulis, the microbiologist, had been around to straighten these guys out. (I used to work in a microbiology lab, too. You learn how to work with pathogens.)

My question to the panel that evening: Why not construct a National Computer Virus Research Lab where there are large networks strictly separated from the outside so that this fundamental work can be done? Russell Brand's answer: "There probably is one already. But because it's dangerous [and socially taboo, I will add] it is therefore secret." The CIA has acknowledged that it has done work with viruses. If the military continues to have the monopoly on computer viral research, then the future of A-life research is in deep trouble.

Richard Thompson showed a really clever computer animation of a self-assembling bacterial flagellum. His color-shaded three-dimensional pieces were the second generation of work started by Narendra Goel at the first A-Life Conference, where Goel used a cheap IBM PC to demonstrate a self-assembling bacteriophage. It starts from a "soup" of randomly mixed parts. Each part has a "configuration bond." This means that certain faces of each part can only bind with certain faces of other parts. For instance, part A may be the shape of a brick, with binding sites on the long sides that can only stick to the short ends of part F. For a bacteriophage there may be dozens of kinds of parts and hundreds of pieces. Double that for the flagellum. You put all the pieces into a virtual bag, and then shake and bake. By giving the pieces certain "physics" and adding the equivalent motion due to thermodynamic energy, the pieces start bumping around into one another. When two matching bonds come close enough, they bind at that joint. Let the mixture run long enough and the structure begins to emerge out of the jumble.

The flagellum of a bacterium is a spinning tail that works like a propeller to drive it through its watery environment. It is, I think, the only example of a wheel and axle in nature. The ingenious method by which the wheel is spun is beyond these notes; it's enough to say that the motor is formed by evenly spaced posts around, but not on, the perimeter of the wheel that "grab" a part of the wheel by chemical bond attraction and move it toward them a minute distance. Collectively, recycling many times a second, they set the wheel spinning as fast as an electric motor.

Thompson was showing this at the A-Life Conference because the means by which even these simple forms are assembled, repaired, and improved without central control is a major question. His work shows that even complicated structures like this can, in theory, arise from some very simple rules. The huge color monitor displayed tiny wedge-shaped nuggets that marched themselves into a helical tube while the base of the wheel and its machinery grew in a strange, half-crystalline, half-embryonic way.

━━

Only a few notes from Norman Packard: "Artificial life is where we will get the answer to, What is life?" "The community is the memory for an organism in a population that is learning." "Rule complexity and high performance go together." "Fluctuation will spawn speciation." "Sex is a computational hassle."

━━

John Holland is another of the founding fathers of A-life. As far as I can tell, he just about invented the field of genetic algorithms. He's an elfish character of indeterminate age who delights in surprising people.

Holland's goal is to design a system that will create complexity from natural selection, rather than from "artificial" selection, as in say, Richard Dawkin's landmark program Biomorphs. In Dawkin's system, the human operator picks out which mutation to breed (much as fanciers breed pigeons or carp), and then his program evolves it. Holland insists that the system itself define the criteria to breed. Or, in other words, that the selection criteria themselves would be an emergent property of that world. (I realize the terminology can be confusing here. Real pigeons are bred with "artificial" selection, while Holland's artificial computer critters will breed with "natural" selection.)

Holland came up with the most sophisticated artificial world I've seen yet. It's a disembodied, pictureless world; everything happens as numbers only without graphic representation. Nonetheless it's come furthest in introducing many of the parameters that ecological systems of life have. Holland calls his world Echo.

In Echo, learning and eating are the same. Echo's creatures live in a grid land, a wide open plane divided into squares. They eat elements. In some squares there are fountains that dish out elements abundantly. Echo's creatures head toward these fountains to consume and be energized. The elements are short bits of code. While the bits of code are food, they also serve as the genes of each individual. (These beings really are what they eat.) For instance, imagine a creature growing genes made up of a's, b's, and c's. In order to use a's in its genes it needs to eat a's. It can get a's from the environment by hanging out at the a's fountain and competing with other organisms for the limited amounts of a's, or it can prey on another organism that has a's in it and eat those, or it can have sex with an organism that has a's in it, swapping needed code. The a's, b's, and c's in a critter's body are added together to form short sequences like genes. The sequence of letters evoke a particular mathematical algorithm, which determines that critter's behavior in seeking out resources. (In many respects this food/gene path is reminiscent of Gerald Joyce's RNA world, where RNA is both the messenger and the message.)

The competition for limited resources, the algorithms that learn over time, and the mutations brought about by sex, all contribute to a wonderfully dynamic ecology in Echo. In just the few short weeks that he has been running the world, Holland has noted some interesting traits.

In Echo, as in Core Wars, the shorter chromosome wins. A short chromosome costs less to reproduce, and is quicker to make, so in a battle, it wins. Recombination (sex) makes an organism a moving target against predators and parasites, and allows it to undergo more change, without as much randomness as simple mutation.

Holland also made an assertion in passing that I believe holds great treasures for biology if A-life can prove it: "Selected mating is the origin of niches."

He is also beginning to track down the food webs produced in these worlds. The consequences of food webs are hardly understood in the wild; having some models for comparison would be explosive. And just for the fun of it, Holland would like to make eggs and seeds, to see what happens. He says he is after "A new mathematics of perpetual novelty. It is this perpetual novelty, and not equilibrium, that equals ecology."

———

Philosopher Robert Rosen preached about the renaissance of biology. He said that biology, through A-life, was going to force physics out of itself. "Whenever biology and physics confront each other, physics has to back down."

———

Rob Collins introduced another ant world. Each ant comprises nine kilobytes in a neural network. There are eight ants per colony, and four thousand colonies in this world. Colonies reproduce, not ants. He ran it with a 0.1 percent mutation rate.

The ants roam the world looking for food, which they are supposed to bring back to their colony nest to fuel reproduction. Like the other examples at the conference, the ants "learn" over generations to better find and compete for food. However, as in real life, individual organisms (which are colonies in the insect world) have their own quirky behavior. Collins found that even though the ants learned to range far for food very early, in some colonies they never learned to pick up food right outside the nest, even after 240 generations. And there was one curious colony that played with their food, stockpiling it in one corner of the world instead of inside their nests. (I don't think they lasted long.) And occasionally some of the ants dropped food into the wrong nest. (This is not uncommon among real social insects, including bees.)

Party small talk in the year 2050: "You can't imagine what my A-life pets did today!"

What I have in my notebook for David Stork's talk is a big half-page diagram illustrating the conference's recurring concept that "there is a price for evolutionary specialization." The chart shows a ball (the generic organism) rolling along a valley road as time goes on. The valley path divides (speciation) into a high road and a low road separated by highlands. Like flowing water, the evolutionary organism tends to seek the lowest energetic route. Once on either road, the organism can't cross over the mountain ridge to change paths. (This depiction is the usual adaptive rugged landscape model inverted. Here optimization is indicated by the deepest valley, rather than the highest peak. In fact, thermodynamics uses the valley metaphor too. The point is the same in both.)

Stork was trying to model the synaptic circuits in the flip of a crayfish tail. This flip is an essential life-or-death escape move for crayfish. Biologically, the circuit is apparently well understood, although how it developed is not. Stork wanted to model the evolution of the circuit from the primitive neural circuits in the surrounding tail muscle which is only used for swimming. How does a very convoluted, tangled, multilevel feedback network arise out of a very plain on/off circuit built for swimming?

Stork used a Connection Machine to simulate a combination of neural equations, genetic algorithms, and learning functions to generate and evolve a population of circuits. Sixty individual circuits competed for 150 generations and were fitness-tested by two simple behavioral tasks, analogous to the reaction needed by a live crayfish. That Darwinian fitness was able to grow successful circuits that mimicked the proper cellular responses was no surprise by now to A-life attendees. What was interesting was that most of the successful circuits had "useless" synapses. That is, the fittest individuals did not always have the optimal circuits. As far as anyone knows these extra links seem like fifth wheels. The conclusions suggest that "as long as the end result and all intervening stages worked, elegance of design system counts for little." This has been a great relief to neurobiologists because they have suspected for a long time that "certain features of the nervous system may not have functional significance."

▭

Przemyslaw Prsuinkiewicz, whose work at the first A-Life Conference was so thrilling, was prevented by pneumonia from giving his presentation this time. Some of his work was shown by associates.

The work he showed at the first conference demonstrated how Lindenmyer Systems (L-systems) could be used to model the shape of plants. Very, very simple formulas could generate forms like ferns and bushes and tree profiles. By layering several L-systems at once, Prsuinkiewicz was able to model the leafing out and flower stages of a wildflower. (It turned out that a return signal from the tips of the flower branch intersecting with up-coming signals from the root would model budding. Botanists are still hunting for chemical evidence of this interaction.) He did this in color. Then he did meadows full of them, each plant beginning as a tiny seed of numbers.

Prsuinkiewicz's current work has taken up Alvy Ray Smith's challenge to re-create the growth of embryos. Together with Martin DeBoer, he has developed recursive rules for the early stages of a single cell dividing into multicell globes with the differentiation of cell types. In particular, they modeled rules that display, say, a color-shape computer graphic of a fern spore growing. There are topological niceties that are accounted for. After a cell divides, the two new halves assume their own optimal shape on the sphere, pushing and jostling neighboring cells to garner the room they want. This constant shuffling of territories is sort of a microscopic plate tectonics. The cell plates snap and bulge somewhat like soap bubbles maximizing the space in a washtub. Even more sophisticated, Prsuinkiewicz used L-systems to simulate the spiral patterns of a growing snail embryo.

The major advance he has contributed to the problem of growing an A-life form is to bring the element of time into the set of growth rules. While his earlier works superbly rendered a bouquet of flowers, or a patch of ferns, they weren't composed the way they actually grew. For example, a branch would be added to the stem, but in reality an embryonic branch develops out of the stem as the stem itself is developing. If the stem is stunted, so will be the branch hidden in it. All growth turns out to be co-growth, just as all evolution is coevolution. Co-growth is what the science of morphogenesis (how things grow) is all about, and why I think that A-life will finally be able to inform the incredible looking-glass world of embryonic organisms, answering such questions as, How does a cell know to become a hippo?

When DeBoer showed Prsuinkiewicz's film of growing A-life plants I was riveted. There was the uncanny recognition of a time-lapse film of real plants surging upward and out, unfolding themselves. His maturing plants gave off an aura that was decidedly organic. There was a grace on that screen that was neither human nor machine.

Thursday night was show-and-tell. Because of the anticipated performance of Survival Research Labs, this session was open to the public of Santa Fe. The entire auditorium was packed. Singles swathed in punk black, deluxe couples out on a date, families with kids. Nothing else going on in Santa Fe that night. Perhaps some were there to find out about artificial life, but most, I gathered, were hoping to see Mark Pauline blow up some robots. There was a lot going in the auditorium, so I'll only get to some of the highlights in no particular order.

Mitchel Resnick from MIT showed his LEGO/LOGO animals. These are neat little toys made up entirely of LEGO blocks. But special LEGO blocks. Certain blocks have little brains in them. These smart LEGOs have chips built into them with metallic contacts for electrical connections. Different blocks have different functions. There are sensor blocks such as the one that senses a wall. There's an "eye" block that detects visible light. One that detects infrared. One that notices whether it's level. Some that hear noise levels. And then there are what one might call logic or cognitive blocks. One is a clock that causes a pause or delay. One called AND gives a signal if a certain stimulus *and* a certain other one is detected. There is one called OR that goes on if one *or* the other stimulus is detected. There's a flipflop that says to do the opposite of whatever was done previously. And of course there are locomotion blocks—little electric motors, gears, and so on—that are off-the-shelf LEGO accessories.

Naturally what you want to do is build creatures with all these blocks, creatures that have the tiniest, dim brain. They do things like follow the light, or run from noises, or run to noises, whatever you cook up. They are for kids and professors.

At the start of their research, the thinking parts of these bricks were inside the blocks. The brains were in an Apple II connected by a bundle of wires to the animals, and the kids could program their behavior using the logic program LOGO (thus LEGO/LOGO). Self-contained as they are now, the MIT folks are calling them Braitenberg Bricks, after Valentino Braitenberg, who proposed similar "vehicles" as autonomous, hard-wired creatures. No, you can't buy them, yet.

<hr />

I think I'll quote from the abstract on the video by Michael Mc-Kenna and David Zeltzer. "The video is about five minutes long. It's an animation called 'Grinning Evil Death'; the story of how a giant six-legged menace from outer space invades the earth and destroys a city. The entire piece is computer-generated; the alien's [a metallic chrome cockroach] motions are automatically generated (as opposed to the more tedious frame-by-frame control methods).

Based on an actual biological model of cockroach locomotion, it adapts dynamically to simulate gravity and obstacles like police cars and power utility lines."

So you have this gigantic chrome cockroach, and all its motions are being pointed to rather than dictated. The programmers say, "walk over those buildings," and the computer-cockroach figures out how the legs go and at what angle the torso should be and then paints a movie of a giant chrome cockroach climbing over buildings. When it jumps down off the other side, a simulated gravity makes its legs bounce and slip realistically.

The programmers considered doing an ant at first, but ants are too cute. This simulated cockroach is swarming with cockroachness. A perfect villain.

━━

It's been a mark of the A-life organizers to keep the boundaries of what is covered in the conferences loose and undefined. This is why Louis Bec, a middle-age French artist, was there. Bec calls himself a "zoosystemician," in the French grand tradition of the naturalist-philosopher-theoretician. Also in the style of the French, he did not give his talk in English.

With the aid of many slides and an emergency French translator, Bec described his fanciful organisms that live in a sulfur world. His pictures showed a loft in Paris that had tons of sulfur piled into yellow soil on the floor. Arising out of this sulfur earth were enormous (up to twenty feet high) beings in fantastical detail. They were invertebrate creatures (stuffed I guess with epoxies and resins), resembling monstrous undersea plants, enlarged microscopic organisms, jellyfishes, slugs, chimeras, internal organs, soft insects, and gigantic molds. Most of all, they reminded me of the unhinged genius found in drawings of an anonymous modern Italian in the book *Codex Seraphinianus*. The resolution of detail was bewitching.

In the installation in Paris, Bec would spend hours upon hours describing to visitors the most minute anatomical detail of his sulfur-based creatures. The walls of the exhibit were covered in scientific charts and tables documenting the natural history of this life, referring in greater and greater microdetail to the digestive tracts, hair follicles, and mate-attracting systems. Every dissection of an internal system was diagramed—how the thing weeps sulfur tears through thin meter-long ducts, and so on. Every part was annotated. Every level was its own wonder. Bec continued, in French, for an hour, taking us on a tour of his artificial-life world. It was a wonderful performance. One had the feeling of being at a lantern show of a visiting nineteenth-century Frenchman just back from a camel expedition to some distant, living part of the universe.

The only presentation to offer criticism of artificial life was a slick video done by Peter Oppenheimer of NYIT. Using computer graphics, he created wonderful lifelike animations, including trees for which he can specify genetic traits. Imagine Pee-Wee Herman's journey into the A-Life Playhouse—big plastic DNA singing songs, daisies with blinking eyeballs, genes with little arms and legs with boots—and you've got the style down. In the same video, Oppenheimer plays the role of Dr. Schizenheimer, who continually raises doubts about his A-life and genetic engineering, only to have himself cloned. To quote from the abstracted plot of this wild video:

> Scientist looking through a microscope sees a universe of micro-organisms. . . . They evolve at the will of the eccentric doctor. By programming his machine he chooses the evolutionary path of the created fantasy organisms. . . . Eventually the doctor looks at the screen and sees himself. He and his clone scream in unison. Now the doctor's choices influence his own evolution. He panics and the machine takes over. He merges with the machine to become a bionic organic live-action animated creature. Finally he falls into a venus flytrap which swallows him. A beautiful graceful orchid emerges. The last scene is of a totally organic world, with no trace of the doctor's technical instruments.

Rudy Rucker demonstrated his commercial cellular-automata program, CA Lab. While investigating different possible rules for his program, he found that the most universal pattern to emerge from many of the rules is a double scroll. This was observed early in CA simulations of a chemical reaction called the Belusoc-Zhabotinsky reaction. You can get a double-scroll Zhabotinsky reaction on a kitchen plate by letting a certain two chemicals mix. When they do, they form these periodically oscillating bands of color on the surface in the shape of double scrolls. Rucker pointed out the similarity of these two-dimensional Zhabotinksy patterns, both in chemistry and in CAs, to the cross-section of fetal embryos, which suggests to him that biological growth may follow the guidance of Zhabotinsky-type reactions.

⊏===⊐

The last ant world shown is actually only one version of a bigger animal world known as Agar: An Animal Construction Kit. Developed by Michael Travers at the MIT Media Lab, the final program will "allow novice programmers to assemble artificial animals from simple components. These components include sensors, muscles, and computational elements. They can also include body parts such as limbs, bones, and joints. A complete animal construction kit will support the coexistence of multiple animals of different species."

As it is now, it is only an ant world. The world Travers showed runs trials of a simulation of cooperative food gathering. Ants are set out into the Agar world. The human zookeeper sets out food anywhere, and the ants will try to find it. When they find food, they lay a chemical trace back to the nest so that the other ants from their nest can find the food more quickly. The chemical trace "evaporates" over time, so sensitivity to the chemical trail is beneficial. The paths the ants take around obstacles are all emergent.

The most important aspect of this project is that the parameters of the environment and the traits of the creatures can be easily modified to produce new creatures and new worlds. Travers is giving away the code to other legitimate researchers interested in improving the A-life in his kit.

⊏===⊐

Brian Yamauchi showed a video of a juggling seeing-eye robot arm that relies on "bottom up" rules. The arm's task is to bounce a hanging balloon on a paddle. This very complex behavior is implemented by a committee of lower "agents" (in Marvin Minsky's terms) that are in charge of a motor or a sensor or another subagent. Rather than have one big brain try to figure out where the balloon is and then move the paddle to the right spot under the balloon and then hit it with the right force, these tasks are decentralized, both in location and in power.

For instance, the problem of "Where is the balloon?" is divided among simple agents, each concerned with the simple question like "Is the balloon anywhere within reach?"—an easier question to act on. The agent in charge of that question doesn't have any idea of when to hit the balloon, or even where the balloon is. Its single job is to tell the arm to back up if the balloon is not within the arm's camera vision, and to keep moving until it is. A network, or society, of very simpleminded decision-making centers like these form an organism that can exhibit remarkable agility and adaptability.

Yamauchi says, "There is no explicit communication between the behavior agents. All communication occurs through observing the effects of actions that other agents have on the external world." Keeping things local like this allows the society to evolve new behavior while avoiding the debilitating explosion in complexity that occurs with hard-wired communication processes. Keeping everybody informed about everything is how intelligence does not happen. Ignorance is sometimes bliss.

It has not been lost on certain astute observers that Yamauchi's recipe is an exact description of a market economy: There is no communication between agents, except that which occurs through observing the effects of actions (note that they see effects but not usually the actions) that other agents have on the common world. This led the Santa Fe Institute (host of the A-Life Conference) to sponsor in 1988 a separate research program on "The Economy as a Adaptive Complex System."

ⅽ══⊐

Belgian scientist Pattie Maes used Rod Brook's six-legged walker (described below) as the experimental animal for teaching a creature how to walk using an agent-based, low-hierarchial system. In this case the thinking for the walking takes place near the two motors for each leg. The leg motors lift or not depending on what the other legs around them are doing. If they can get the sequence right ("Okay, hup! One, three, six, two, five, four!") walking "happens." As I understand it, the sequence is another job set by an agent. Getting up and over obstacles like a mound of phone directories required adding some sensing whiskers to send ground information to the first set of legs. Since the other legs are watching the first legs, walking over obstacles happens. There is no one place in the contraption where walking is governed. There is no way for a motor, say, to determine whether it is walking or not. It knows only if it is moving its legs up and down. Sometimes when they are moving up and down the creature is stuck. Sometimes if everything is in harmony, the creature walks. But the parts don't know.

One of the major principles to be elucidated at the first A-Life Conference was the thrilling notion that complex behavior in a variety of systems from computer grid worlds to biological immunities, from synthetic ecologies to global economies could all be produced with what are called "local rules." Local rules guide the behavior of individual agents. These bottom-up heuristics say nothing *directly* about what happens at further levels. If birds on the fly keep a certain distance between neighboring birds on the fly (a local rule), then they will exhibit a characteristic flocking behavior (a global rule), depending on what local rules they start with. Therefore, flocking (a global behavior) emerges out of local behavior. You can't get flocking by having each bird keep in mind the shape of the flock and try to do its part to keep it that way. Such a strategy would be too error-prone even if it were possible.

Visualizing levels of emergent order that originate at the localest grass roots rules and cascade up, yielding self-organization at increasing scales of complexity, is the easy part. One can intuitively see how, with a clever choice of laws, local rules can govern global behavior. The hard part is understanding how global behavior can govern local rules.

The agent of the paddle doesn't know where the balloon is; the walking motor doesn't know if it is walking; the bird on the fly doesn't know the shape of the flock. Yet it became apparent at the second A-Life Conference that it wasn't as simple as that. The kind of perceptual novelty that John Holland talks about arises when there is a return communication between local and global. Somehow the global must control the local, difficult as it is to hit a moving target. Somehow the flock can aim itself toward a destination, and sometimes over years, the flock will change its destination, or even, by evolution, what the aim of flocking is. All these changes entail the locally elected global order governing the local. Start with simple rules, get complex behavior—easy part. Get complex behavior to govern simple rules—very hard part. This is an important and final loop in a very recursive circuit. How this loop stays flexible, rather than becoming an ever-tightening noose, is, I bet, the theme of the third A-Life Conference.

Ian Galton made the mathematical case that both fractals and L-systems converge on the same attractors, making them a subset of what Alvy Ray Smith calls "graphtals." I'm calling them "growthals."

Tom Ray presented Tierra, an artificial life simulator. Written in assembly language, it is a free-for-all core war, somewhat in the manner of Rasmussen's VENUS world. It uses the metaphor that CPU time is "energy" and memory is "materials." Organisms must execute certain instructions without error, or else be terminated by the "reaper," a routine that withholds memory space from the stupid. Only the fittest survive. (I wonder how long it will be before we see "highly evolved" as a computer advertisement cliche?) Ray reported both a pattern of punctuated equilibrium over time as well as the speeding up of adaption by parasites, confirming earlier experiments by others in the room.

MaClennan began with an apt quotation from Karl Popper, written in 1974: "The main task of human knowledge is to understand it as continuous with animal knowledge; and to understand also its discontinuity. . . ."

His main concern was with the creation of synthetic ethology, the study of synthetic animals, or what he calls "simorgs." He kept using the word "game" and I realized that throughout the conference, game was a word that came up again and again. To create a fitness function, points are scored; creatures move around on a grid, like a board game; events proceed in unison by "turns"; there are rules. Games have always led computers, from chess to pong. It is very probable that artificial life will be a compelling (addictive I say) set of computer games played intimately and expertly by prepubescent kids before it ever gets to the ethics committees of science funders.

A-life will be treated as a toy. It will enter our lives through the young, which will ensure its future.

Richard Belew's talk was about evolving networks. He said he was interested in the coupling between learning and evolution. Some of the things he noted from his experiments, which use genetic algorithms to design and train networks, were in agreement with others. He too found that genetic algorithms could construct unanticipated solutions to problems. He also said something that sounded very profound on first hearing, but I haven't been able to test it yet. "Adaptation means: to capture regularities over a range of time scales."

Games again. Stephanie Forrest has been doing pioneer work in game theory. She worked with Axlerod on the classic Prisoner's Dilemma, and has recently been applying genetic algorithms to solving arms-race problems, and "nonlinear international relations." Using a model that parallels John Holland's work, she has been looking at how simulated countries can evolve their negotiation strategies for mutual benefit. In Holland's A-life world each organism determines its strength of defense, offense, reproduction; in Forrest's nonlinear international world, each country determines its strength of guns, butter, reinvestment. Her initial results imply that in three-country worlds, the strongest position arises when the two weaker countries join as allies.

What is of interest here is that the allegiances are formed without prior leading. It is an emergent property of a complex system. What the Prisoner's Dilemma and Forrest's current work point to is "how cooperative behavior can arise in populations of autonomous self-interested agents in which there is no central authority." In worlds that seem to be propelled by Darwinian-described competition, how does cooperation ever arise? Richard Dawkins's explanation is that selfish genes make altruistic organisms. Forrest is suggesting other ways.

Remarkably it is the ants, again, who have some answers. Other than A-life fans and their ant worlds, the only other set of people seriously investigating the question of emergent cooperation are the real-ant fans—the sociobiologists. Ants exhibit rule-breaking altruism where one would not expect it—in pretty dumb and savage little beasties. They have systematic cooperation that is not corrupted by individual little-mindedness, and this is of interest to political scientists as we try to restructure a global economy. The Book of Proverbs (6:6) speaks truthfully when it says, "Consider the ant's ways and be wise!" Ants are, it appears now, the world's leading experts in nonlinear international relations.

For this one, you had to be there. David Ackley had misinterpreted the conference's request for video demos, and instead of slapdashedly copying some last-minute computer screens onto a cassette, he produced a gonzo, informative, and hilarious tape of the best talk during the whole week. He would have had a standing ovation if anyone had had any strength to stand up by this time. He did get a thundering round of applause and cheers. I and many others pressed him to make his video available commercially. It's the one I would recommend as the best initiation for those who have no inkling of what A-life is.

Ackley has a screen presence like David Letterman. (I must confess that I've only seen this video once, after a long week of incredibly poor communicators and lousy audiovisuals. Maybe I wouldn't have found it so brilliant if I saw it at the beginning of the week, or outside the conference.) In the video he invites us, the audience, to look over his shoulder as he explains his very graphic A-life world. His creatures have human faces. (No ants for him!) These humanoids run around in his world trying to acquire the usual things—resources, energy, and right answers. They bump into walls if they are not careful. They are winnowed out if they are wrong and don't get smarter. They have genes that guide their behavior, and they undergo mutations and reproduce sexually. They breed faster than rabbits. An all-nighter on Ackley's computer may take them to three hundred generations.

As others have, Ackley found that his world was able to evolve amazingly fit organisms. Successful individuals would live Methuselahian lifetimes (twenty-five thousand day-steps in his world)—virtual immortality. These creatures had the system all figured out. They knew how to get what they needed with minimum effort, and how to stay out of trouble. Not only would individuals live long, but the populations that shared their genes would live long as well.

Noodling around with the genes of these streetwise creatures, Ackley discovered that he could make some improvements in their chromosomes that would make them even better adapted to the environment he had set up for them. He discovered a couple of ways to exploit resources that they hadn't taken up. So in perhaps the first act of virtual genetic engineering, he modified their evolved code and set them into his world. As individuals, they were superbly fitted and flourished easily, scoring higher on the fitness scale than any creatures before them.

But Ackley noticed that their population numbers were always lower than the naturally evolved organisms. As a group they were anemic. Although they never died out, they were always within the range of an endangered species. Ackley felt that if he ran his world for more than 300 generations, they might not last. So while the handcrafted genes suited individuals ideally, they lacked the robustness of organically made genes, which suited the species ideally. Here, in a lab, in the home-brewed world of a midnight hacker, was the first bit of testable proof for hoary ecological wisdom: that what is best for an individual isn't necessarily best for the species.

"It's tough accepting that we can't figure out what's best in the long run," Ackley said, "but, hey, that's life!"

—————

Alan Kay's presentation was quite disappointing. He flew in Friday morning having missed the week, and having thus missed the audience. He gave his usual stuff about kids ("Making large concepts accessible to children is the best way to make it accessible to ourselves"), but he should have simply let the audience ask him questions. In 1982 he tried to answer the question, "How would you program a virtual aquarium?" His just-do-it answer—the Vivarium project—started out as a kit for making virtual fish. He said they soon found out that more than a kit, they wanted an environment for virtual life. That would have been an ideal place to open the discussion to the floor. Next time.

—————

John Nagle talked about squirrels. Taking a cue from Hans Moravec at the CMU Robotics Lab, who suggests that current computers have the intelligence level of a snail, Nagle argued for aiming at the realistic goal of generating the intelligence of a squirrel. Rodents have about one gram of brain mass, which Nagle says is equivalent to a computer running somewhere between 100 and 1,000 MIPS (millions of instructions per second). That level is not as impossible as human-level AI, but far more useful than ant-level intelligence (your average Macintosh).

Squirrel-level intelligence will get us automatic character animation. Specifically, he proposed a goal of reaching such a level that an A-life squirrel "does the right thing over periods of less than one minute." He said the hard things that A-life and AI folks tend to "abstract out" in order not to do, are actually the most important things. Going along with the Moravec graph of increase in computer power as a function of time, Nagle said we'll have squirrel-level artificial intelligence by the year 2000.

A young woman (one of the few present) stood up at the microphone at question time and told Nagle that he was full of it. She said that almost all the scientists dreaming about A-life anytime soon were on cloud nine, that none of them had any idea of how complicated real biology was, that Moravec was out to lunch, that the work on retinas that he has been basing his projections on is shoddy, that she is a neurobiologist herself who happens to be studying the retina and Moravec hasn't got a clue as to how computationally sophisticated the eye alone is, that computer scientists like him underestimate the power of animal brains to the point of silliness, that equating MIPS with intelligence is one good example of how crippled the whole A-life movement is, and would he care to comment?

Nagle shifted uneasily and said he had a plane to catch (he did) and maybe someone else could answer her (they didn't).

At the first A-life Conference there was a 4-H contest for the best A-life creatures. This time there were few entries and the prizes were given somewhat cursorily. I can't even remember who won. But I do know who should have won. Without deliberation, I would have given the blue ribbon to Rod Brooks's six-legged insect robot.

Brooks runs the robot lab at MIT. He says that rather than try to bring as much life into A-life as possible, he's trying to bring as much A-life into life as possible. He wants to flood the world (and beyond) with inexpensive, small, ubiquitous thinking things. He's been making robots that weigh less than 10 pounds. The six-legged walker weighs only 3.6 pounds. It's constructed from model car parts. In three years he'd like to have a one-millimeter (pencil point–size) robot. He has plans to invade the moon with a fleet of shoebox-size robots that can be launched from throwaway rockets. It's the ant strategy. Send an army of dispensable, limited agents coordinated on a task, and set them loose. Some will die, most will work, something will get done. In the time it takes to argue about one big sucker, Brooks can have his invasion built and delivered. The motto: Fast, Cheap, and Out of Control.

Fast, cheap, and out of control robots are ideal for:

- Planet exploration
- Collection, mining, harvesting
- Guiding
- Remote construction, say of a lunar base

As an example he built what he cheerfully calls the Collection Machine, a robot in an office space that collects empty soda cans at night. It's ingenious. It operates according to the society-of-mind approach to A-life robotics. The eyes of the Collection Machine spot a soda can on a desk and guide the robot until it is right in front of the can. The arm of the robot knows that it is in front of a soda can because it "looks" at its wheels and says, "My wheels aren't turning, I must be in front of a soda can." Then it reaches out to pick the can up. If it is heavier than an empty can, it leaves it on the desk. When it takes a can it finds its way all the way back to its station to unload it, then randomly wanders again through offices until it spots another can. (A variation, called the Confection Machine, dispenses candy to people in exchange for their opening doors for it.) Not very efficient per trip, but night after night it can amass a great collection of aluminum. During the day it sleeps.

Brooks has another small robot in mind that lives in the corner of your living room or under the sofa and wanders around vacuuming at random whenever you aren't home. You only notice how clean the floor is. A similar, but very tiny, insectlike robot lives in one corner of your TV screen and eats off the dust when the TV isn't on. A student of his built a cheap, rabbit-size robot that watches where you are in a room and calibrates your stereo so it is perfectly adjusted as you move around.

Brooks's most ambitious plan is to send a flock of tiny solar-powered bulldozers to the moon five years in advance of a proposed lunar base program. They can be built from off-the-shelf parts in two years, and launched completely assembled in the cheapest possible one-shot lunar-orbit rocket. Operating entirely by "local rules," without any communication from earth, they will daily scrape away soil needed to level building sites. When the expedition arrives, the astronauts will turn the robots off and give them a pat.

Fast, Cheap, and Out of Control.

===

I'm relating Chuck Taylor's closing summary of the conference here, before my report of the last panel that preceded his summary, because the panel discussion was the more fitting end, and in fact Taylor wrote his conclusions before hearing the final panel anyway.

A-life, according to Taylor, will be the research that will contribute to "a unified theory of nature." He listed six specific goals that the talks of the week indicated would become important.

- To understand why complexity increases
- To assist in software and wetware engineering. If we can't build it, we can evolve it
- To prepare us for the dynamics of the future
- To understand what "life" means
- To aid in the study of specific organisms, ecologies, and the origins of biological life
- To advance the reintegration of biology back into art and culture

===

The last panel consisted of Alan Kay, Mark Pauline, Chris Langton, Doyne Farmer, Rod Brooks, Rudy Rucker, and Norm Packard jawboning about "what is the future of A-life?"

At some point Mark Pauline finally got his shock-wave cannon working. On the evening before during the A-life show-and-tell, he had cranked up his cannon using an electric hoist. The blue warning light had blinked on ("Cover your ears when you see the blue light go on"), but nothing had happened. After several rounds of this blue light, no shock, they determined that a solenoid on the cannon had broken on the plane flight here. The crowds had to be content to watch videos of past Survival Research Laboratories spectacles. They were awesome and disturbing. One avant-garde video was a stark "documentary" of elaborate dinosaurlike machines involved in ceremonial rituals of sacrificing other machines to the machine god. These were not sleek Star Trek machines, but rusty, smoldering, smoky, greasy, vibrating, mechanical monsters. They have gears and pistons, and sharp edges. Other videos showed some of their performances in years past ("Illusions of Shameless Abundance," "Failure to Discriminate," etc.)—modern-day Roman circuses of mechanical gladiators done under searchlights, screeching loudspeakers, the crackle of fire, and the smell of diesel oil.

The idea is to do the shows with as little publicity or official approval as possible, have the audience get as close as they dare, and make the machines as fast, cheap, and out of control as one can. For instance, take the "Flame Thrower on Wheels." It used a Mack truck V8 engine to run a huge blower that sucked up kerosene from a fifty-five gallon drum and ignited it with a carbon arc, spewing out a tongue of vicious orange flame easily a hundred feet long. It was controlled by a little model airplane joystick. Or Pauline's description of his newest pet: "This completed device is based on electromagnetic rail-gun technology. Rather than firing a projectile at high speeds for kinetic impact effects similar to other droll, destructive military objectives, this device employs similar capacitor energy-storage units to liquify a metal bar and magnetically eject the molten blob at about two hundred miles per hour. It appears to the eye as a cometlike beam that fragments on impact, tending to set fire to any nearby combustibles." Pauline added, "This machine is SRL's answer to George Bush's call for a thousand points of light."

Nothing as kinetic was planned for the A-Life Conference. What was hoisted in the air, and not working the evening before, was a mere shock-wave cannon. At some point in the afternoon of the last day, Pauline got it working. A tube about five feet long is chained to an extended platform near the speakers' podium, and pointed over the heads of the 200 conferences. Hoses run to a tank of acetylene on the floor. The blue light buzzes on. Boooooooooooom. A terrific crack blasts the hall, blowing off all the papers and notes and handouts on the tables of the participants in the back row. The scientists scatter. A rain of dust and plaster bits showers down. The blue light goes on again. Boooooooooooom. I am standing behind the cannon and actually see a wave go across the conference hall and hit the back balcony wall, where it shatters the plaster stucco. People are ducking now. Another blue light. Again, Booooooooooom. It's actually less noise than a thump to the chest. Pauline is enjoying it, his impassive face with a hint of a smile.

During the panel Alan Kay proposed that the ancient Greek dictum, "the visual arts imitate the art of creation," has been modified now to "the computer arts imitate creation itself."

Langton said he is learning that in artificial life, the part that is artificial is not the life, but the materials. It is real life in an artificial medium. He is one of many who sees humans and their machines as part of the natural evolution of life.

Norm Packard made the observation that in A-life, preserving harmony is more important than preserving species. Something he said later prompted me to note that one thing involved in A-life that few researchers have mentioned is time compression. These models compress evolutionary time into human scale. This, of course, accelerates the rate of change in evolution. But, more important, I think A-life will continue to accelerate the differences in time cycles that technology has introduced into the world. With the creation of A-life we will have formal hierarchies of time. Whether the slow will govern the fast, or vice versa, isn't known, but the control of velocity will lie near the heart.

Brooks called for an infiltration of robots. He's been working on seeing how "dumb" you can make a robot and still have it do useful work. He gave the example of smart doors. For only about ten dollars extra you could put a chip brain in a door so that it would know you were about to go out, or it could hear from another smart door that you are coming, or it could notify the light that you left, and so on. If you had a building full of these smart doors talking to each other, they could help control the climate, as well as help traffic flow. If you extend that to all kinds of other apparatus we now think of as inert, putting fast, cheap intelligence into them, then we would have a colony of sentient entities, serving us, and learning how to serve us better.

His prediction for the future of A-life is that we'll have creatures living with us in mutual dependence—A-life symbiosis. They will be small, ubiquitous, hidden, and taken for granted. Their numbers will outnumber us, as do insects. And in fact, his vision of robots is less that they will be R2D2s serving us beers, than that they will be an ecology of unnamed things just out of sight, engineered with an insect approach to problems—many hands make light work. Ant World approacheth.

━━

"The movie *Frankenstein* is an albatross around the neck of A-life."

—Doyne Farmer

"But the book is great. It should be required for all A-life studies."

—Chris Langton

━━

Mark Pauline: "Machines have something to say to us. When I start designing a SRL show, I ask myself, what do these machines want to do? You know, I see this old backhoe that some redneck is running everyday, maybe digging ditches out in the sun for the phone company. That backhoe is bored. It's ailing and dirty. We're coming along and asking it what it wants to do. Maybe it wants to be in our show. We go around and rescue machines that have been abandoned, or even dismembered. So we have to ask ourselves, what do these machines really want to do, what do they want to wear? So we think about color coordination, and lighting. Our shows are not for humans, they are for machines. We don't ask how machines are going to entertain us. We ask, how can we entertain them? That's what our shows are, entertainment for machines."

━━

"Yeah, when machines are superintelligent and superefficient, what is the niche of humans? Do we want machines, or do we want us?" asked Langton.

Pauline responded, "Humans will accumulate artificial abilities, while machines accumulate biological intelligence. This will make the confrontation even less decisive and less morally clear."

━━━

Somebody on the panel, at the very end: "You know Rod Brooks's smart doors? Eventually we won't say to the door, 'Open'; we'll have to say, 'Open, please.' "

━━━

Chris Langton closes. "Okay, everybody, thank you for coming. It was wonderful. Mark, how about one last time for the future of A-life?"

The blue light goes on.

Boooooooooooooooooooooooooom!

What Are the Laws of Nature?

PAUL DAVIES

A schism comes . . . when considering the status of law-like statements. Are they to be regarded as discoveries about reality or merely as the clever inventions of scientists? Is Newton's inverse-square law of gravity, to take a famous example, a discovery about the real world that happens to have been made by Newton, or is it an invention of Newton's made in an attempt to describe observed regularities? Put differently, did Newton uncover something objectively real about the world, or did he merely invent a mathematical model of a part of the world that just happens to be rather useful in describing it?

Once upon a time, in the Age of Magic, people believed in gods to make things happen. There were rain gods, sun gods, tree gods, and many more. As knowledge about the natural world grew, these gods were no longer necessary and were abolished, until there remained only the one true and ultimate God who designed, created, and supervised the operation of the universe.

The ancient Greek philosophers, while remaining within a generally theistic paradigm, recognized that the physical world displayed certain systematic and regular features. The Pythagoreans discovered the role of number in nature, and the mathematical concept of ratio gave us the word *rational*. The Judeo-Christian tradition affirmed the existence of a real external world that is distinct from, but was created by, God. What is now called science is in fact a distinctive form of Western rationalism that owes its conceptual structure to the legacies of Greek, Judeo-Christian, and Muslim philosophies. The early stages of this science retained strong theological roots, developing the view that the world could be understood by rational inquiry and must therefore be the manifestation of a rational deity. Thus the scientists of the sixteenth and seventeenth centuries proclaimed that God invented mathematical laws of nature to make His creation run smoothly.

The idea that natural phenomena conform to mathematically precise laws received a huge impetus with the work of Issac Newton in the seventeenth century. Newton formulated laws of motion and gravitation that were taken to be universal. In the centuries that have followed, Newton's laws have been replaced by others. Notwithstanding the changing fashion regarding their precise form, the belief that mathematical laws of some sort underpin the operation of the physical world is now a central tenet of scientific faith. Indeed, it is hard to imagine what we could mean by science if there were no such laws.

Although the conceptual foundations of modern science owe their origin to theological worldviews, the theistic dimension has faded. The laws that were once regarded as a manifestation of a rational creator (thoughts in the mind of God) have become "freestanding." Thus the laws have taken on the status formerly reserved for God and are imbued with the same mystical properties: They are universal, eternal, absolute, transcendent, omnipotent, and so forth. So central to science have they become that they are taken for granted, without need for further explanation. Yet I cannot myself leave it at that. I want to know what these laws are. Where do they come from? Do they have to have the form that they do?

Such questions, although they deal with the very essence of science, are not usually regarded as scientific. Indeed, most scientists do not trouble to ask them at all. Yet recent progress at the forefront of physics has revived the question of what meaning should be attached to the laws of nature, and the differing philosophical stances adopted by scientists has led to sharp differences in their approach to research. I have been astonished in conversations with my colleagues by how divergent and woolly their thinking is on this most fundamental of subjects. To the layperson, it may appear surprising that something as basic as the nature of physical laws remains contentious among professional physicists—but it does.

What is not at issue is the fact that the workings of nature exhibit striking regularities. The orbits of the planets, for example, are described by simple geometrical shapes and display distinct rhythms. Patterns and rhythms are also found within atoms and their constituents. The concept of clockwork attests to the mathematical precision evident in the motion of mechanical structures. These are examples of what, at rock bottom, might be called the dependability of nature: the fact that the world goes on existing and behaving in much the same way from day to day.

On the basis of such experiences, scientists have used inductive reasoning to argue that these regularities are lawlike. If the apple falls today, it will do so tomorrow. The attraction between a north and south magnetic pole will not suddenly become a repulsion.

Water will always freeze at zero centigrade. And so on. The philosopher David Hume demonstrated that, in fact, we have absolutely no right to reason this way. There is no basis in logic for your supposing that just because the sun has risen every day of your life, it will also rise tomorrow. The belief that it will—that there are, indeed, dependable regularities of nature—is an act of faith, but one that is indispensable to the progress of science.

A key element in the scientific worldview is the separation of the laws governing a physical system from the states of that system. The state of a physical system is defined by specifying the values of certain parameters, such as pressure and temperature or the position and momentum of all the constituent particles. The state is not something fixed and God-given; it will generally change with time. By contrast, the laws, which provide correlations between states at subsequent moments, do not change with time. They are eternal. And in this respect they are, of course, a vestige of the timelesss, eternal nature of the Judeo-Christian God, to whom appeal was originally made as a source of these laws.

A related fact concerns our own relationship with the physical world. The state of the world is *malleable;* we can change it. Indeed, it is central to the methodology of empirical science that human beings can prepare well-defined physical states and study their evolution, and it is the business of technology to gain control over the workings of nature and manipulate them for our purposes. However, we cannot change the *laws:* these are "God-given" and eternal. Our science and our technology must remain forever confined within the constraints of those laws.

The separation of laws and states is reflected in the mathematical structures employed to model the physical world. In classical mechanics, for example, the dynamical laws are embodied in a mathematical object called the Hamiltonian, which acts in phase space. On the other hand, the state of the system is represented by a point in phase space, and this point moves about with time. The essential fact is that the Hamiltonian and the phase space itself are *independent* of the motion of the representative point.

This much is agreed. A schism comes, however, when considering the status of lawlike statements. Are they to be regarded as discoveries about reality or merely as the clever inventions of scientists? Is Newton's inverse-square law of gravity, to take a famous example, a discovery about the real world that happens to have been made by Newton, or is it an invention of Newton's made in an attempt to describe observed regularities? Put differently, did Newton uncover something objectively real about the world, or did he merely invent a mathematical model of a part of the world that just happens to be rather useful in describing it?

The language that is used to discuss the operation of Newton's laws reflects a strong prejudice for the former position. Physicists talk about planets' "obeying" Newton's laws, as though a planet were inherently a rebellious entity that would run amok if not for the fact that it is "subject" to the laws. This gives the impression that the laws are somehow "out there," lying in wait, transcending the actual hardware of the physical world, ready to supervise the motions of planets whenever and wherever they occur. Falling into the habit of this description, it is easy to attribute an independent status to the laws. But is this really justified?

How can the separate, transcendent, existence of laws be established? If laws manifest themselves only through physical systems—in the way that physical systems behave—we can never get "behind" the stuff of the cosmos to the laws as such. The laws are *in* the behavior of physical things; they are not free-floating. We observe the things, not the laws. But if we can never get a handle on the laws except through their manifestation in physical phenomena, what right have we to attribute to them an independent existence? Is this not to fall into the well-known trap of confusing the map with the territory or the model with the reality? (I am aware of the claims that mystics can apprehend the reality directly, but that is another story!)

A helpful analogy here is with the concepts of hardware and software in computing. The laws of physics correspond to software, while the physical states are the hardware. (Granted, this stretches the use of the word *hardware* quite a bit, as the definition of the physical universe includes nebulous quantum fields and even spacetime itself.) The foregoing issue can then be stated thus: Is there an independently existing "cosmic software"—a computer program for a universe—encapsulating all the necessary laws? Can this software exist without the hardware?

There is a parallel conundrum concerning the status of mathematics. The statement "Eleven is a prime number" is indisputably true. But was it true before the existence of mathematicians? Presumably so, for the quality of primeness is timeless. If eleven is prime today, it must always have been prime and always will be. The same goes for Pythagoras' theorem: It was a true statement long before Pythagoras. This seems to suggest that mathematical statements, theorems, and laws have an independent existence, that mathematics is "out there," in some abstract realm that transcends space and time, and that mathematicians discover already existing mathematical relationships as their investigations become ever more refined.

According to this philosophy, there are, at this time, many true theorems still waiting to be discovered by future generations of mathematicians. There are also statements, such as Fermat's famous last theorem, that are either true or false, whose truthness or falsity we do not yet know, though we may well do so soon. Such statements, it is said, *really are* either true or false, whether we know it or not. Furthermore, the logician Kurt Gödel showed that there are mathematical statements that are true but cannot be *proved* true, implying to some that there is indeed much mathematics "out there," beyond the grasp of mathematicians.

Plato, who proclaimed the existence of eternal Ideas that transcend space and time, accepted the foregoing philosophy, and it was adopted either explicitly or implicitly by many mathematicians and scientists in the centuries that followed. Rudy Rucker, for example, quotes Gödel as defending Platonism by saying, "I do objective mathematics."[1] With the discovery that the laws of physics can be cast in mathematical form, it became natural to suppose that these laws, like the mathematics that expressed them, enjoy an independent, absolute, eternal, and transcendent existence.

There is no doubt that the mathematization of physics has contributed greatly to the Platonic view of the laws of physics. Physicists are often at pains to point out the inadequacy of everyday commonsense language for describing, say, the weird world of quantum physics or relativity. This can be captured only mathematically. Indeed, some physicists have adopted the extreme position that the world *is* mathematics and nothing more. "God is a mathematician," proclaimed James Jeans, and many a modern theoretical physicist would agree.

But whatever the status of mathematics vis-à-vis the laws of physics, not all mathematicians agree that mathematics is "out there" anyway. The opposing view is that mathematicians do not discover mathematics, they invent it. The concept of primeness, for example, is said to be merely a definition dreamed up by mathematicians for their own purposes. Mathematical relationships are then simply tautologies, akin to the relationships between characters in a novel or a play. Thus the statement "Eleven is a prime number" has a similar status to the statement "Romeo is Julie's lover." The latter statement is true, but it would be considered absurd to suppose that it was some timeless absolute truth.

If this position is correct, there really are no independently existing laws of physics at all. What we sometimes call laws are merely a way of systematically ordering the facts of our experience into a coherent worldview that happens to be currently fashionable. What we call "the" laws are better described as "our" laws; they are not "right" or "wrong," but only more or less useful to us within the context of our current paradigms.

What do the physicists think? Arthur Eddington pondered deeply the nature of physical law. He distinguished what he called identical laws from transcendental laws. The former, among which he included laws such as the conversion of energy, he regarded as "imposed by the mind." They are "obeyed as mathematical identities in virtue of the way in which the quantities obeying them are built. They cannot be regarded as genuine laws of control." On the other hand, "If there are any genuine laws of control of the physical world, they must be sought in . . . the transcendental laws" for which "we are no longer engaged in recovering from Nature what we ourselves put into Nature, but are at last confronted with its own system of government."[2]

This distinction between laws that are merely human inventions and "genuine" transcendental laws is stated more succinctly by physicist Howard Pattee. "I picture the universe as the primitive concept, the primitive elements of the universe being external to subjects or living systems and being governed by physical laws. Now, I don't mean by physical law *our* description. I mean something outside—the real thing. And I say these laws are *inexorable, universal,* and *incorporeal.*"[3]

Richard Feynman also believed that certain fundamental laws of physics have independent existence. He liked to draw the analogy between the laws of physics and the rules of chess. Physicists, he said, are like spectators of a chess game, deducing existing eternal rules by watching how the pieces move. Shortly before Feynman died, I had occasion to interview him for the BBC. I asked him directly whether he thought of the laws of physics as existing in their own right, independently of the universe. He began by discussing mathematical relationships. "When you discover these things, you get the feeling they were true before you found them. So you get the idea that they existed somewhere, but there's nowhere for such things," he said. "Well, in the case of physics we have double trouble. We come upon these mathematical interrelationships, but they apply to the universe, so the problem of where they are is doubly confusing. . . . I get the feeling that I'm discovering laws that are *out there,* analogous to the feeling that a mathematician gets when he discovers laws that are *out there.*"[4]

The assumption of eternal laws, independently existing "out there" but as yet only dimly glimpsed by man, permeates much of the thinking of physics engaged in the recent ambitious attempts at unification.[5] These so-called Theories of Everything have as their goal the amalgamation of all fundamental particles and forces, and even space and time, in a single mathematical scheme. Various unified field theories and, more recently, theories that aim to build the world out of strings[6] all make the assumption that there is a single fundamental dynamical principle controlling all physical processes, waiting (somewhere in the wings) to be elucidated. Physics thus becomes a search for the "right" principle. In practice this has tended to reduce to a search for a mathematical entity known as a

Lagrangian (closely related to the already mentioned Hamiltonian), from which certain dynamical equations may be recovered by using a well-defined mathematical procedure.

To be sure, existing formulations are conceded to be flawed; currently fashionable Lagrangians are only approximations to "the correct Lagrangian of the world." The assumption is that as the unification theories progress, they will more and more closely image "the real Theory of Everything," which by implication must somehow already exist abstractly. The history of science is thus seen as a sequence of successively better approximations, or fits, of "reality." In his much quoted inaugural lecture to the Lucasian Chair in Cambridge, Stephen Hawking declared that "the end of theoretical physics" is in sight. Hawking envisaged as imminent the correct identification of "the" Lagrangian, thus making physics a closed subject.

Ultimately, any unification program has to tackle the age-old problem of the origin of the universe. For the greater part of scientific history, the coming into being of the universe as a whole has lain beyond the scope of science altogether, but in recent years there have been a number of attempts to encompass "the creation" within the scientific framework generally and within some sort of unified theory in particular. Most cosmologists now accept that the universe has not always existed, and that its appearance was an abrupt affair that occurred about 15 billion years ago—the so-called Big Bang. It is essential to realize that the Big Bang represented the coming into being not merely of matter and energy, but of space and time, too. Thus it is meaningless to ask what took place before the Big Bang, because there was no before. In the words of St. Augustine, "The world was made with time and not in time."

This raises pressing questions for the notion of eternal laws, for if the laws of physics came into being *with* the universe, one cannot appeal to those laws to explain the creation event. If we are to explain how the universe came to exist "from nothing" (not even from empty space), then the laws that control the physical processes that brought about its appearance must transcend the universe they create. They must somehow already exist. Heinz Pagels summed up the issue poetically as follows:

The nothingness "before" the creation of the universe is the most complete void that we can imagine—no space, time, or matter existed. . . . Yet this unthinkable void converts itself into a plenum of existence—a necessary consequence of physical laws. Where are these laws written into that void? What "tells" the void that it is pregnant with a possible universe? It would seem that even the void is subject to a law, a logic that exists prior to space and time.[7]

Attempts to tackle the creation of the universe from nothing have been pursued largely within the context of what is known as quantum cosmology. The essence of quantum physics is indeterminism: Quantum events occur without well-defined causes. If quantum mechanics is applied to the universe as a whole, it provides a possible theoretical framework for describing a universe that comes into being spontaneously, as a "quantum fluctuation from nothing."

Any such attempt has to confront the question of initial conditions. As formulated, all known dynamical laws refer to classes of processes, rather than to individual systems. To take a simple example, it is a law that the paths of all pitched baseballs are parabolas, but the *particular* parabola of a *particular* pitched ball is not completely determined by this law. There can be long, shallow parabolas; short, high parabolas; and so on. The actual shape of the path adopted by a given ball will depend both on the general law and on the specific initial conditions—the speed and angle of pitch. Given the law plus initial conditions, the path is uniquely determined.

In familiar systems, the initial conditions are fixed in some way by the wider environment; in the case of baseballs, by the pitcher. But when it comes to cosmology—the dynamics of the entire universe—there is no wider environment. Therefore, the initial cosmological conditions cannot be fixed by usual physical criteria. They can be retroactively inferred, of course, from the present state of the universe, but not explained by appeal to other physical processes. The question of why the Big Bang went bang in precisely the form it did—or, for that matter, why it went bang at all—remains a mystery. As a result, one might be tempted to regard the cosmological initial conditions as outside the scope of science, perhaps (like the laws?) as God-given.

An alternative position has been enunciated by James Hartle and Stephen Hawking.[8] They propose a "law of initial conditions," formulated within the framework of quantum cosmology, that fixes the precise way in which the Big Bang went bang. This law is really a mathematical prescription (or ansatz) that selects the quantum state of the universe uniquely. As with any law, it is proposed as a hypothesis, to be tested by observation; there is no suggestion that the Hartle-Hawking ansatz is the only conceivable prescription. The authors do, however, argue that it is in some sense the most natural. Moreover, the selected state is entirely consistent with the currently observed large-scale structure of the universe.

Now, it goes without saying that the Hartle-Hawking law must be regarded as transcendent. Because it fixes the state in which the universe starts out, it cannot be a law that comes into operation only once it has started out. Moreover, this law differs fundamentally from all others in two ways. First, it applies not to a class of systems, but to a single system—the universe as a whole. Second, it is timeless, though not in the sense that it applies unchanged throughout time. Rather, it is relevant to the beginning of time only.

To physicists engaged in the unification program, or in discussing deep questions of creation and existence, the laws of physics seem almost as real as the universe itself. The opposing idea that the laws, or the superlaw that will one day unify them, are somehow mere conventions, or projections of the human mind, appears ridiculous. These theorists are convinced that they are dealing with reality.

The notion of physical reality has undergone a subtle yet profound change with the rise of the "new physics." As already remarked. Western science initially adopted the Judeo-Christian picture of an independently real external world created by God and subject to eternal laws, and then dropped God. In classical physics, the reality is in the *state* of the world—in the hardware. The laws are shadowy abstract entities that constrain, but do not constitute, that reality. In the new physics, however, this position cannot be sustained. First, the theory of relativity undermines the concept of "the" state, because there is no universal time at which to specify the state. The state of the world as judged by a given observer depends on the motion of that observer. And to connect the states associated with different observers, one must use the laws.

Second, in quantum physics the state is not real anyway. What evolves with time, according to the dynamical laws, is not "how the world is," but an abstract mathematical object that encodes the relative probabilities of multiple contending realities. It is more an expression, as Niels Bohr put it, of "what we can say about the world," a sort of amalgam of observer and observed, or of hardware and software. Where, then, is the reality in quantum physics? Some physicists (John Wheeler, for instance) insist it is in our observations alone. Quantum cosmologists tend to eschew this view and tacitly (I contend) inevitably fall back on the position that the reality lies in the "Hilbert space" and the Hamiltonian for the system—roughly speaking, in the laws. Thus, the classical position is turned on its head. The laws (Hamiltonian, Hilbert space) are regarded as *objectively real,* while the state of the system is regarded as (at least partially) subjective.

The strong belief in the objective reality of the laws is reinforced by what has been called the unreasonable success of mathematics in physics, a sentiment echoed by many theoretical physicists today. After all, it is perfectly possible to imagine a world in which material objects and natural phenomena are not ordered according to simple mathematical relationships. Yet experience reveals that, as Roger Bacon expressed it, "Mathematics is the door and the key to the sciences. . . . For the things of this world cannot be made known without a knowledge of mathematics."[9] From the inverse-square law of gravitation to the abstract gauge groups of modern unified field theories, mathematics is the language that most succinctly encapsulates the workings of nature.

Some philosophers, such as Kant, have claimed that we invent precisely the mathematics that enables us to make sense of the world. It is then no surprise that we find nature conforming to simple mathematical laws. But physicists are suspicious of this argument. Very often mathematicians elaborate whole branches of mathematics entirely for their own interest, and only later does this mathematics unexpectedly find application to the real world. Furthermore, the *same* types of mathematical relationships that crop up in one area of science often turn out to be equally fruitful in other, completely unrelated, and at the time undiscovered, areas.

It is also odd, to say the least, that the deeper we probe into nature, the more convincingly does our mathematics work. The investigation of the structure of matter and the forces that control it provides a striking example. As physicists peer deeper and deeper into matter, the power of mathematics to unify and order the subnuclear world becomes more manifest. This escalating power of mathematics to unify in an elegant and compelling way is highly suggestive of a convergence toward a unique mathematical statement—"a formula that you can wear on your T-shirt," to use Leon Lederman's words—that will explain all of physical reality in a single, embracing scheme.

The fact that mathematics works so well in fundamental particle physics, where the ultimate building blocks of matter are exposed, thus convinces most theorists that nature is in some sense founded upon objective, eternal, transcendent mathematical principles. If mathematics were merely invented by humans to make sense of the world, we would have no right to expect it to work so well at the fundamental level. Indeed, we would expect it to be most useful at the level of everyday experiences. But in fact, the powerful mathematical relationships of fundamental physics are masked in the human realm by all sorts of complexities.

It would be quite wrong, however, to give the impression that belief in the transcendence of physical laws is the orthodox position among scientists. In my experience, scientists who work on "applied" problems tend to believe that the laws of nature are just convenient models that happen to be useful for ordering natural phenomena but do not enjoy an independent existence. Thus an electrical engineer would argue that Ohm's law, which relates electrical resistance to current and voltage, was invented to give a simple description of electrical circuits. And it certainly appears odd to suggest that Ohm's law is a transcendent absolute truth that already existed and was somehow lying in wait for billions of years until somebody built an electrical circuit.

Physicists usually distinguish between *fundamental* and *phenome-*

nological laws. The distinction is, however, rather subtle and intuitive. Quantum electrodynamics, or QED, which describes the interaction of electrons and photons at the microscopic level, involves fundamental laws, but Ohm's law is regarded as phenomenological. The difference stems from the fact that QED is a completely general theory, applying to electrons and photons under essentially all circumstances, whereas Ohm's law refers to a very specific and complicated physical system (an electrical circuit).

It is often claimed that phenomenological laws can in principle be reduced to, or explained by, fundamental laws—that, for example, Ohm's law might be derivable, ultimately, from QED. In practice this can almost never be achieved. Nor is the claim strictly correct as it stands. True, Ohm's law is consistent with QED, and QED is almost certainly *necessary* for the validity of Ohm's law; but it is not *sufficient*. In addition to QED, one needs to make all sorts of additional assumptions about constraints, boundary conditions, and statistics that are not logically part of QED at all.

This distinction is crucial, I believe, when one confronts the puzzle of whether the laws have independent existence. Earlier I used a computer analogy: Can the software exist independently of the hardware? Well, one must be mindful that the "hardware" of physics is not simply given once and for all but evolves with time. Some physical systems that exist today simply were not to be found in the universe at much earlier epochs. A good example concerns the phenomenon of superconductivity, the complete loss of resistance to electric current that occurs in some materials at very low temperatures. Superconductors can be made in the laboratory and their properties studied. Physicists have discovered that they obey certain definite mathematical laws. Did these laws in any sense exist before the advent of superconductors?

The point is that superconductivity may not occur outside the laboratory. (There is some evidence that it may occur within neutron stars, but quite probably for many millions of years prior to the formation of stars, there were no superconductors anywhere in the universe.) Therefore, does it make sense to think that the laws of superconductivity existed in the primeval universe, waiting, so to speak, for the first superconductor to come along? I think most physicists would answer no to this question. But then it would be equally absurd to suppose that a new law suddenly comes into existence, and instantly propagates throughout the cosmos, when the relevant hardware appears.

The fact that the universe as a whole evolves, and that the hardware of the cosmos is not a fixture, has encouraged the speculation that the laws of physics might also evolve with time. Dirac proposed that the strength of the gravitational force gradually fades, although his formula for this has been ruled out by observation. However, this is not really what I mean by an evolving law, because Dirac merely replaced the usual law of gravity with another law that described how the first law systematically changed.

A more profound idea stems from the discovery that at higher and higher energies (or temperatures), laws that were once considered inviolable might in fact be violated. For example, it was discovered long ago that the law of conservation of matter is violated when new particles and antiparticles are created in high-energy processes; more recently, according to certain unified field theories, that the law of baryon conservation (stating that the total number of protons and related particles must remain constant) might fail slightly at ultra-high energies. It has been suggested that if one could drive the energy up without limit, one by one all of our cherished laws would fail, leaving only chaos.[10] These laws would be seen not as eternal transcendent truths, but as incidental low-energy constraints on the behavior of matter.

In practice we cannot drive up the energy (or temperature) of the universe without limit, but in the context of the Big Bang theory this happens naturally. The creation event itself can be regarded as a state of essentially infinite temperature. According to the chaos theory, then, the well-ordered universe we see is a frozen relic of primeval chaos. There was no lawfully supervised coming into being out of nothing. Everything came, to use the words of John Wheeler, "out of higgledy-piggledy."

Wheeler has elevated what he calls the fundamental *mutability* of the physical world to the status of a principle—in fact, the *only* principle. "The only law," writes Wheeler, "is that there is no law."[11] He has been led to this extreme position by the study of gravitational collapse, which predicts the existence of so-called spacetime singularities. These are boundaries to spacetime where all physics seemingly breaks down. Wheeler favors a cosmology in which the entire universe eventually collapses in a "big crunch," like the Big Bang in reverse, culminating in an all-encompassing singularity. Thus, space and time, as well as all matter and energy and all physical structures, would be destroyed. Reasoning that all laws of physics are formulated within the context of spacetime, Wheeler concludes that no law can survive the crisis of gravitational collapse.

If the physical world is not, after all, underpinned by eternal laws, where does its apparent lawfulness originate? Wheeler has an astonishing proposal, a sort of hyper-Kantian extreme. We the observers create that lawfulness by our very observations. Appealing to the oft-cited melding of observer and observed that emerges from quantum physics (and which I shall discuss shortly), he ventures, "Could it be that the *observership* of quantum mechanics is the ultimate underpinning of the laws of physics?"[12] In this "participatory universe" we are the lawmakers, not by passively projecting our mental forms onto the physical world, but actively, *by making observations that shape that world into lawfulness.*

The idea that the laws of nature have somehow evolved out of lawlessness is not new to philosophy. Friedrich Nietzsche conjectured that "the origin of the mechanical world would be a lawless game."[13] But just as biological order arises out of molecular chaos, so might the mechanical world "ultimately acquire such consistency as the organic laws seem to have now." Moreover, the existing laws would have emerged by a sort of natural selection from among a vast set of alternatives: "All our mechanical laws would not be eternal, but evolved, and would have survived innumerable alternative mechanical laws." The neuroscientist Richard Gregory recently expressed to me a similar belief, that the laws of physics have arisen from a type of Darwinian selection.

The difficulty with this idea is that selection is meaningful only in the context of an ensemble of identical systems. Selection can occur in a biological species, for example, but what meaning has it in the case of a unique universe? One possibility is to conjecture the existence of an ensemble of universes. Those with "well-adapted" laws will "prosper" in the sense of evolving in an orderly, interesting way to states of greater and greater organization and complexity until they spawn conscious beings who write articles about the meaning of it all. The universes less well served by their laws will degenerate into chaos and anarchy, and go unobserved.

The selection mechanism here is known as the anthropic principle, and it has been much discussed in recent years.[14] It implies that the universe we live in has its distinct lawlike properties not because these laws are eternal and transcendent, but because we have selected such a universe by our very presence. What we mistakenly call "the" universe is in fact but one among a vast ensemble of generally uninhabitable universes. The lawfulness of our universe is then no surprise, for lawfulness is a prerequisite of habitability and hence cognizability.

Another imaginative proposal has been put forward by Rupert Sheldrake.[15] The controversial biologist, who believes that there are no eternal laws, instead draws an intriguing comparison between the traditional timeless laws of physics and statute law. But what, asks Sheldrake, if the laws of nature are more like common law? In other words, what we take to be absolutely inviolable, transcendent truths are, in his view, really trends, or habits, accumulating over the eons out of an initially lawless state by repetitive application and mutual reinforcement.

If one accepts the idea of evolving laws, the distinction between fundamental and phenomenological laws fades away. As new and ever more elaborate physical systems come into existence with time, so might the laws that apply to them. Thus one can regard life as an emergent quality, and the concomitant laws of biology as emergent alongside it. What were once regarded as primary laws, such as Ohm's law or the laws of superconductivity, become so by temporal priority alone. The laws of mechanics were there first because isolated particles predated electric circuits. But the former would not be more fundamental than the latter, at least in the deep sense considered hitherto.

If I interpret him correctly, this seems to be the position of Ilya Prigogine, who, when asked whether the laws of nature can undergo transition, replied: "The laws of nature may indeed depend on the state of nature. . . . Can you speak of laws of biology if there is no life? Obviously not. In other words, what we call laws or regularities depend on their realization."[16] And elsewhere he stated: "There is no longer a single fundamental level. Our level, the level of macroscopic beings, is not less fundamental or more fundamental than the microscopic level . . . there is no longer an absolute level of description. There are various levels, all interconnected in a much more complex fashion."[17]

Prigogine's suggestion that the laws of nature may depend on the state challenges the basic assumption that runs through all of orthodox science: the conceptual separation between eternal laws and the state of the physical system. As far as we know, this separation of "software" (laws) from "hardware" (actual physical state) is justified for all simple systems. But I have a sneaking feeling it may fail for complex systems. I have argued elsewhere that the act of observation in quantum mechanics entails a "hardware/software entanglement."[18] The biologist Robert Rosen has developed an elaborate mathematical description of complex systems in which there is no separation of dynamical laws and states.[19] If one were to pursue this as far as cosmology, it would imply that the universe is a sort of gigantic feedback system where the laws that apply today depend on the details of cosmic history.

These evolutionist ideas are unlikely to find favor among particle physicists, cosmologists, and unifiers, whose thinking is strongly influenced by the theory of relativity. Central to the evolutionary theme is the dimension of time. The laws of nature, like the cosmos itself, are given a history. *An arrow of time is built into the operation of physical laws.* Time is thus singled out from spacetime and ascribed a metaphysical significance that it lacks in relativity theory, where space and time are on an equal footing and spacetime itself is part of the dynamical system. Indeed, Hawking and co-workers are busy deemphasizing the significance of time, even to the extent of trying to redefine it as space in disguise.[20]

Let me turn to the question of whether the laws in any sense *have* to be what they are. Could the world have been otherwise? And if so, how has a selection been made? It has long been a dream of the unifiers that when we finally write down that elusive Theory of Everything in a single magnificent formula (that you can wear on your T-shirt, remember), this superlaw will be the only mathematically and logically self-consistent statement. Thus one reads the following in the book *Gravitation,* by Misner, Thorne, and Wheeler: "Some principle uniquely right and uniquely simple must, when one knows it, be also so compelling that it is clear the universe is built, and must be built, in such and such a way, and that it could not possibly be otherwise."[21] It is this issue that I believe Einstein was referring to when he said that what really interested him was whether God had any choice in the construction of the world.

If the remarkable position were reached that *unique* statements of both the dynamical laws and the law of initial conditions could be explicitly displayed, then the only question left would be why there is something rather than nothing. Note, however, that because of quantum uncertainty, even the complete circumscription of the laws and initial conditions in the manner described would not suffice to fix everything about the world. It would almost certainly not fix you and me, for example. (Presumably, the superlaw would also demonstrate why this same uncertainty is an indispensable ingredient in the logical self-consistency of the world.)

The above claim is very hard to argue convincingly. For a start, it glosses over all manner of difficulties of a Gödellian nature concerning the shaky foundations of logic and mathematics. The concept of "the only logically and mathematically self-consistent system of laws" is almost certainly either meaningless or undecidable in principle. Furthermore, it is actually rather easy to think of alternative universes to the real one that are not obviously mathematically or logically *inconsistent*. For example, there seems to be nothing impossible about a Newtonian world of classical nonrelativistic rigid particles, perhaps interacting also by electromagnetic fields based on an ether.

Of course, it might well be possible to establish the much weaker result that there is only one set of laws (or superlaw) consistent with the existence of conscious organisms. This would explain why we observe the laws that we do. They would be not the only possible laws, but the only ones we could possibly observe. That does not actually explain why the laws are as they are, though. They *could* have been otherwise, and then the universe would have gone unobserved. Our universe would be unique in being "aware of itself."

Some selective mechanism then has to be invoked. This could be a good old-fashioned deity (God selected from a large, perhaps infinite, set of alternative laws just that unique set that would grace his creation with conscious beings to contemplate his actions). Alternatively, *we* could be the selectors. I have already briefly mentioned this type of anthropic reasoning in connection with any sort of lawfulness. There could be an ensemble of universes, each with different laws, and only in ours would the laws permit biology and hence observers. Or one could argue that observers are actually indispensable to the logical self-consistency of the world, so that the laws we observe (or something very like them) are indeed the only possible set.

Finally, let me declare my own hand. I have always taken a somewhat positivist approach to the nature of reality. It seems to me that, ultimately, we have to work entirely with the (common) facts of experience. Physicists are in the business of making models to relate these facts. In the course of this endeavor, they invent certain concepts, such as energy and atom. At rock bottom, these concepts are merely code words that encapsulate certain complicated mathematical properties of these models. However, they become so familiar to us that we tend to promote them to the status of real, independently existing entities. When we do that, we try to imagine them "out there," possessing certain objectively real qualities. We then run into the trouble that these hypothetical things simply *cannot* be imagined. Whatever mental image you may have of an atom, it is wrong. Certainly it is not a little solar system of orbiting particles, nor is it a packet of pulsating waves. It is nothing that we can visualize.

What, then, is an atom? I would say that it is an abstract concept—a Platonic Idea, if you like—that helps us relate certain types of observations in a systematic way. "Atomic theory'" is actually simply a set of algorithms for effecting these relations (or more accurately, explicating correlations), and *atom* codifies the greater content of these algorithms. The atom is an abstract entity that belongs to the model; it does not really exist as an objective, independent object.

The reader might be concluding that I am denying the reality of the external world. What I am denying, however, is the adjective *external*. In modern physics, subject and object are indivisible. The "real world" consists not so much of "things out there" observed by us, but of *connections*—specifically, connections between observations. The reality is the connections. John Wheeler makes this point in characteristically flamboyant form: "What we call reality, that vision of the universe that is so vivid in our minds, we plaster in between a few iron posts of observation by an elaborate labor of imagination and theory." He summarizes this in a pithy quotation from Torny Segerstedt: "Reality is theory."[22]

You might also believe that I am subscribing to the "cultural" viewpoint, by arguing that atoms, and the laws of atomic physics, belong only to our models and not to the "real world." Well, I agree that there is an element of truth in this, because our models are undeniably products of our cultural milieu. But it is not the whole truth. Unless one is prepared to retreat into the lonely fastness of solipsism, one has to concede that the discoveries of science reveal that there is *something* orderly about the world, something not totally the result of our imaginings. The very fact that our observations can be fitted into mathematical models and algorithms surely tells us something about reality that we had no right to expect a priori.

Let me try to make this more precise. When we say that there are laws of nature, we mean that physical systems share certain properties and patterns of behavior wherever and whenever they are. Thus, the motion of an electron in a hydrogen atom on earth is the same as that on a distant quasar and was the same billions of years ago as it is today (there is good observational evidence for this). Likewise, the paths of falling bodies on earth and on a far-flung galaxy belong to the same geometrical class.

When you stop to think about it, this coherence across space and time is a surprising and mysterious fact. It is tempting to ask, "How does one electron (or falling body) know what the other is "doing?" The orthodox answer is that it doesn't; both comply with the laws of physics. But this must mean that these laws have some sort of objective existence that transcends space and time. Or so it seems to me. There is another way of expressing this, using the concept of computer algorithms. If we could take a "God's-eye view" and perceive the whole of space and time at once, together with the intricate tracery that is the collective history of all the particles of matter, we should notice that this texture is not random but patterned. The paths traced out in spacetime (what physicists call world lines) are correlated and ordered and have a preponderence of particular shapes. One could give an account of this panorama by dividing spacetime into minute cells and using binary arithmetic to register whether a cell was traversed by a particle path or not. The result would be a colossal array of zeros and ones, representing the total information content of the history of the universe.

It is then meaningful to consider a computer algorithm that would generate the same sequence of zeros and ones. If the algorithm had the same information content as the sequence it generated, one would say the information was *incompressible*. This would be the case if the universe were completely random. In fact, the cosmic information is highly compressible. (When a computer is used to predict, say, eclipses, it is utilizing—on a very small scale—precisely the compressibility of cosmic information.) This compressibility we codify into "laws of physics," which certainly have a cultural content. But the fact of compressibility is not explained by our culture. (I should mention, as an aside, that the information content of the world, while enormously compressible, is not quite as compressible as was once thought, owing to the existence of so-called chaotic systems.)

So I find myself uneasily poised between two contending factions. On the one hand I am convinced that there are *some* transcendent eternal truths that constrain the nature of the physical world, and I believe that the coming into being of the universe was a lawful event and not the genesis of law. However, I am suspicious that what are conventionally regarded as "fundamental laws" (for instance, the Lagrangian of some collection of fields or particles or strings) are altogether too narrow to encompass the full richness of physical phenomena, especially the organizational properties of complex systems. Thus I see the need for laws at all levels of description and dislike the application of the designation "fundamental" to the bottom level alone. I therefore agree wholeheartedly with Anthony Leggett when he wonders whether "the mere presence of complexity or organization or some related quality introduces the new physical laws" and goes on to suggest that "quantum mechanics breaks down in the face of increasing complexity."[23] Thus I believe that new laws come into play as the universe evolves to ever more complex states.

Can one have one's cake and eat it? Is it possible that the emergence of "laws of complexity" is itself lawlike in the eternal sense? In other words, might it be that there exist transcendent principles of a mathematical-logical nature that endow the universe with a predilection for certain laws, but the laws are not themselves eternal? This could be a neat route between the horns of the age-old dilemma concerning creativity in nature.

The problem of creativity is this. We recognize that as the universe evolves, new forms and structures come into being; this is especially obvious in the biological realm. How are we to explain this creative power of nature? Plato supposed that the appearance of a new form merely made manifest the already existing transcendent Idea of that form. So the new form was already latent in earlier stages of the universe, a "phantom awaiting its hour," as Henri Bergson so beautifully put it. The same essential concept resurfaced with the Newtonian mechanistic universe and Laplace's famous claim that all future states of the universe are completely determined by its present state plus the eternal laws of motion. Thus the future is already contained in the present; indeed, everything has been fixed in detail since time immemorial. The appearance of a giraffe, say, is not a genuinely new event because it was already implicit in all previous states of the universe.

The problem with the foregoing causal explanation is it tells us merely that things are the way they are because they were the way they were. We are left to explain where the giraffe idea came from in the first place (or alternatively, why the cosmic initial conditions contained states that will one day develop into giraffes). The other option is to accept an element of spontaneity in nature and put the appearance of the giraffe down to pure chance. This means abandoning the Newtonian-Laplacian mechanistic world view, which is, in any case, discredited nowadays (because of quantum mechanics, for one thing). However, attributing something to pure chance alone in no sense explains it. On the contrary, doing so is to assert that it has no explanation.

The conventional Neo-Darwinian account of the giraffe augments chance with another principle: natural selection. Thus the giraffe *is* explained, but not in the causal sense of Newtonian mechanism. The explanation instead involves a higher-level organizing principle, a "law of complexity." But this law, or principle, of natural selection did not come into existence until the formation of the biosphere. So where, exactly, did *it* come from? My suggestion is that such higher-level laws of complexity as the principle of natural selection are possible, but not inevitable, consequences of the eternal laws of physics. They cannot be reduced to those laws because they depend on the state of the world as well, and we now know that this state is not completely determined by the laws plus initial conditions.

Thus the eternal laws endow the universe with a *predisposition* to develop along certain evolutionary pathways (I believe, for example, that they make the appearance of life somewhere in the universe more or less inevitable), but they do not fix in detail everything that will come to pass. In particular, they do not fix giraffes. So an element of spontaneity is retained within a generally causal-deterministic framework. There exists in the eternal laws (including maybe a law of initial conditions) something like a Platonic Idea, or blueprint, for a universe of roughly the sort we observe, and in particular a universe of growing complexity. As states of ever-greater complexity emerge spontaneously, new higher-level organizing laws and principles emerge with them. So it is that the physical world has given rise to conscious beings who observe the world about them and ask: What are the laws of physics?

NOTES

[1] Rudy Rucker, *Infinity and the Mind* (Boston: Birkhäuser, 1982), 168.

[2] A. S. Eddington, *The Nature of the Physical World* (Cambridge: Cambridge University Press, 1930), 244–245.

[3] In Paul Buckley and F. David Peat, eds., *A Question of Physics* (London: Routledge & Kegan Paul, 1979), 120.

[4] Transcribed in P. C. W. Davies and J. Brown, eds., *Superstrings: A Theory of Everything?* (Cambridge: Cambridge University Press, 1988), 207–208.

[5] For a nonspecialist review, see Paul Davies, *Superforce* (New York: Simon & Schuster; London: Heinemann, 1984).

[6] For a nonspecialist review, see Davies and Brown, op. cit.

[7] Heinz Pagels, *Perfect Symmetry* (New York: Simon & Schuster, 1985), 347.

[8] Stephen Hawking, *A Brief History of Time* (New York: Bantam, 1988).

[9] Roger Bacon, *Opus Majus*, trans. Robert Belle Burke (Philadelphia: University of Pennsylvania Press, 1928).

[10] H. B. Nielsen, "Field Theories Without Fundamental Gauge Symmetries," in W. H. McCrea and M. J. Rees, eds., *The Constants of Physics* (London: The Royal Society, 1983).

[11] J. A. Wheeler, "Beyond the Black Hole," in H. Woolf, ed., *Some Strangeness in the Proportion* (Reading, Mass.: Addison-Wesley, 1980).

[12] J. A. Wheeler, "Genesis and Observership," in Butts and Hintikka, eds., *Foundational Problems in the Special Sciences* (Dordrecht, Holland: Reidel, 1977).

[13] "Eternal Recurrence: The Doctrine Expounded and Substantiated," in O. Levy, ed., *The Complete Works of Friedrich Nietzsche,* vol. 16 (Henley-on-Thames: Foulis, 1911).

[14] See, for example, P. C. W. Davies, *The Accidental Universe* (Cambridge: Cambridge University Press, 1982) and John Barrow and Frank Tipler, *The Cosmological Anthropic Principle* (Oxford: Oxford University Press, 1986).

[15] Rupert Sheldrake. *The Presence of the Past* (London: Collins, 1988).

[16] Renée Weber. *Dialogues with Scientists and Sages: The Search for Unity* (London: Routledge & Kegan Paul, 1986), 189.

[17] Paul Buckley and F. David Peat, eds., *A Question of Physics* (London: Routledge & Kegan Paul, 1979), 74.

[18] Paul Davies, *The Cosmic Blueprint* (New York: Simon & Schuster; London: Heinemann, 1987), ch. 12.

[19] For a reasonably nontechnical review, see "Some Epistemological Issues

in Physics and Biology," in B. J. Hiley and F. D. Peat, eds., *Quantum Implications: Essays in Honour of David Bohm* (London: Routledge & Kegan Paul, 1987).

[20] Hawking, op. cit.

[21] C. W. Misner, K. S. Thorne, and J. A. Wheeler, *Gravitation* (San Francisco: Freeman, 1973), 1208.

[22] Wheeler, "Beyond the Black Hole."

[23] A. J. Leggett. *The Problems of Physics* (Oxford: Oxford University Press, 1987), 177–178.

Exploring Future Technologies

K. ERIC DREXLER

Nanotechnology will be based on molecular machines and molecular electronic devices. With computers and robotic arms smaller than a living cell, it will enable the construction of almost anything, building up structures atom by atom.

Everyone knows that technology has reshaped our world. Advances in technology have transformed a world of isolated peasants toiling for a handful of lords into a world where ever more millions of people are free to roam a globe echoing with satellite newscasts. Most of us suspect that advances are far from over. Fewer have tried to understand future breakthroughs. Many would call the effort futile.

Yet if understanding is possible, it seems worth seeking. If the gross trends in technology are any guide to the future, we will face dramatic advances in a time frame for which we pretend to make serious plans.

Bureaucrats blithely project Social Security budgets for the year 2025, as if they know what another thirty years' progress in computers, artificial intelligence, and robotics will mean for economic productivity—and as if they know what another thirty years' progress in the biotechnology revolution will mean for health care and life expectancy. To assume no change in these fields would be ridiculous, and to assume that the future will *merely* bring longer lives and better factories seems an exercise in fantasy.

Parents plan college for newborn infants, implicitly planning career preparations for a stranger in a strange world. They save for their own retirements, and a personal journey into that world. If we make long-term plans for saving and spending, then surely it makes sense to consider developments that will determine what our money can buy—if, that is, there is anything of importance we can puzzle out. My purpose here is to show how one can puzzle out something of importance about future technologies, and to sketch a few of the larger results.

These results, if correct, should affect public policy decisions. It seems that certain lines of work will lead to technologies of great

75

power. If so, then—given standard policy assumptions, in a competitive world—it makes sense to pursue those technologies, to try to gain their benefits while forestalling their abuse. Research-and-development budgets in the United States alone total tens of billions of dollars annually, and long-term prospects affect short-term research goals. With a better understanding of the future, we could invest more wisely today.

===

To think productively about future technologies is largely a matter of *exploratory engineering*. To do a better job of understanding the future, we need to do a better job of understanding, practicing, and judging efforts in exploratory engineering. The following sections of this essay examine this field, comparing it to science and standard engineering.

Exploratory engineering involves designing things that we can't yet build. This may seem a dubious proposition. "If we can't build something, who will pay for its design? Surely, the designs will be sketchy and inadequate. And, if we can't build it, how will we test it? Surely, any conclusions about its workability will be tentative and inadequate." These are natural questions, and the answers to them revolve around the counterquestion, "Inadequate for what?"

One shouldn't expect the exploratory engineer to concoct specific designs and propose them as the definitive machines for the 2006 model year. Not only are present engineers too ignorant to do so (a basic problem), but we lack the resources. Modern industrial designs are often complex and sophisticated; future designs will likely be more so. On a small budget, one could not possibly design today's machines, much less the future's.

So why bother with exploratory engineering? Because, just as one can have a general knowledge of today's machines and what they can do—without knowing their detailed designs—so one can, perhaps, gain a general knowledge of some of tomorrow's machines. A general knowledge can include important facts, and detailed, sophisticated designs are not essential to a general understanding. It is one thing to know about cars, roads, petroleum, and suburbs; it is another thing to know about the complexities of internal combustion engines and how these affect car weights and gas mileage.

Likewise, consider nuclear bombs. The key points in a general understanding of them are simple: They work by nuclear reactions, initiated by fission, releasing nuclear levels of energy and producing active nuclear debris. This general understanding (and more) was possible before the Manhattan Project, and hence before the first bomb. Sophistication was secondary. The first crude bombs were grossly inefficient by today's standards, yet they beat conventional explosives by orders of magnitude. Even the still cruder exploratory designs for these bombs must have shown the possibility of this, because the basic potential lay not in the sophistication of the designs but in the fundamental principles of the technology.

The lesson in this is simple: In new technologies of fundamental power, even clumsy, conservative designs can sometimes give awesome performances. Exploratory engineering works best in these new domains, where crude designs can beat the most sophisticated systems possible with present technology. The chief example explored here is a domain called *nanotechnology*, a technology based on molecular machines able to build molecular machines—and anything else. Here again, crude, exploratory designs are enough to reveal great power.

By its very nature, however, exploratory engineering has nothing to say about the timing of events. There is nothing in the design of a machine that tells how long a community of human beings will take to develop all the technologies needed to build it, or whether they will try. Dates do not fall out of design calculations.

These uncertainties limit the value of exploratory engineering, if one seeks predictions of future events rather than the directions of future progress. One might guess at matters of timing, but it is wise to be cautious in these guesses. What "cautious" means, of course, depends on the issue at hand: A technophile's optimism is a technophobe's pessimism, and vice versa. If one considers the unprecedented economic and health care benefits promised by nanotechnology, the cautious assumption is that it will take a long, long time to arrive. If one considers the unprecedented potential for abuse of nanotechnology, however, the cautious assumption is that it will arrive with startling speed. For those chiefly concerned with the direction of progress—with choosing productive lines of research, for example—uncertainties about the timing of long-term goals are less important.

━━

Exploratory engineering, more than most engineering, builds on science—yet this does not make it a branch of science, any more than bridge design is a branch of science. And this is important to recognize, because the confusion between science and engineering is fatal to understanding the future of technology. To judge by newspaper and television coverage, space flight is a great achievement of science, and scientists spend a lot of time trying to make rocket engines work. Any scientist or engineer, of course, will tell you that space flight—though it has benefited from science and in turn yielded scientific knowledge—is itself an achievement not of science but of engineering. Understanding the difference between these fields is vital: If engineering were a science, then exploratory engineering would be impossible.

Science and engineering build on each other and use similar tools, but they have different goals. Science aims to understand how things work; engineering aims to make things work. Science takes the thing as given and studies its behavior; engineering takes a behavior as given and studies how to make something that will act that way.

This difference makes foresight impossible regarding scientific knowledge, but not regarding engineering ability. The limit on foresight regarding knowledge is simple and logical: If one were to *know* today what one will "discover" tomorrow, it wouldn't be a discovery. Since engineering is about doing rather than discovering, no such logical problems arises. There is no contradiction in saying, "We know that we will be able to land a man on the moon," as Kennedy's advisers did in the early 1960s. When scientists do make predictions about their future knowledge, they predict what they will learn *about* rather than what they will learn. And this is often a matter of engineering: "We will learn about the composition of the lunar surface—because engineering will take us there."

Because of this situation, confusion about science and engineering hinders understanding of future technologies. If we confuse engineering with science, then we will think that little can be said about its future—that engineering *projections* are as poorly founded as scientific *speculations*. And we will tend to think that scientists (with their proper and ingrained distrust of speculation) are the right experts to ask about the future of technology. Scientists have little reason to ponder the nature of engineering and often misunderstand how it differs.

"Thus, we conclude that engineers can, through the discipline of exploratory engineering, tell us solid facts about future technological prospects. We need only ask them, and listen to their answers." This is, of course, nonsense.

Standard engineering has a short-term perspective for a simple reason: Employers will not pay engineers to think about what can be built in another fifty years, because there is no money in it. In the United States, companies seldom pay engineers to think about what can be built in ten years. Medium- and long-term exploratory engineering are little practiced today. What is more, the discipline of exploratory engineering differs so greatly from that of standard engineering that standard engineers may be excused for doubting whether it even makes sense.

═══

Engineering is about designing things—ordinarily, things that one can build, test, and redesign in the short term. Exploratory engineering is about designing things that can be built, but only with tools that we don't yet have; this makes it a different sort of endeavor.

The differences begin with motive. Standard engineering receives massive funding to help achieve a competitive advantage in the world—to build a more attractive CD player or a more aggressive jet fighter, and to do it soon. Exploratory engineering, to the extent that it is practiced at all, seeks to construct not a physical artifact but a rough understanding of future technological capabilities.

The exploratory engineer must still do design work of some sort, or there would be nothing to discuss, no real ideas to criticize. But those designs may omit many details and still make a solid case for a future capability. In standard engineering, the job isn't done until every detail is specified, since every detail must be built. This makes the exploratory engineer's job simpler.

Since exploratory engineering aims to build a solid case rather than to outperform the best similar system, it need not try to push the limits of the possible. This has profound consequences for the nature of the intellectual enterprise. Again, it makes work simpler.

In standard engineering one seeks a net advantage in any way that works, regardless of whether we understand *why* it works. Engineers must seek lower-cost manufacturing, which forces them to work with all the complexity of factory operations. They must seek better materials, which drives them to confront all the complexity of metallurgy and polymer chemistry, as in manufacturing turbine blades for jet engines and composite materials for wings. They may have to push the limits of precision, cleanliness, purity, and complexity, as in state-of-the-art microprocessor production. And almost any production process is likely to use a big bag of tested, reproducible black-magic tricks: Add a pinch of this, a dash of that, and clean the glass with Alconox detergent before step five.

Though engineers eagerly use (and produce) scientific knowledge, they no more *need* to understand how a process works than a bird needs to understand aerodynamics. Cut-and-try works in engineering as it works in other evolutionary systems. Once discovered, a process may work, prove its reliability in testing, and provide a real competitive advantage—yet it may remain utterly beyond analysis and simulation based on current knowledge. Competitive pressures encourage engineers to increase their understanding, but those same pressures do not allow them the intellectual luxury of staying on well-understood ground.

Competition pushes engineering beyond the limits of what can be analyzed or simulated. In practice, the tools of analysis and simulation alone never yield competitive systems and hence can never support the full burden of standard engineering. This requires testing, not only to get the details right, but to discover valuable yet mysterious processes. But exploratory engineering must rely almost exclusively on these limited tools: One can't test and learn from what one can't build. The standard engineer, looking at this situation, has the strong gut feeling that analysis and simulation will prove inadequate—as indeed they would, if the goals and designs were those of standard engineering. But when a design need not be competitive, then—at least in some instances—it need not go beyond what can be analyzed in terms of well-understood laws of nature.

In these instances, analysis and simulation can give strong reason to think that a rough, exploratory design could in fact be made to work. To make such a case, the designer must pay attention (explicitly or implicitly) to a host of questions. What are the relationships among materials' properties, component shapes, strengths, forces, speeds, energies, temperatures, voltages, currents, chemical reactions, and so forth? The list is long, but (for any particular class of physical system) it is still finite.

━━

Uncertainties in analysis and simulation pose problems. Different fields suffer different problems from uncertainty; once again, confusion about the differences separating science, standard engineering, and exploratory engineering can confuse our efforts to understand the future of technology.

In exploratory engineering, one can't test and measure, so uncertainties may remain large. Some can be dealt with by leaving large margins for error in a design. If you don't know how strong the material will be, assume the worst and beef up the thickness of the part to match. Where a standard engineer would be forced by competitive pressures to leave only an adequate margin for safety (after testing to probe the limits of workability), the exploratory engineer can design in a huge margin for ignorance, just to make a more solid case.

In an unknown environment, uncertainties may also be unknown. Accordingly, exploratory engineering is easier when the designer can assume a simple or well-defined environment. For parts inside a machine, the machine is the environment and is itself a known part of the design.

A more subtle problem arises when a precise result must follow from a combination of uncertain quantities. For example, a spinning part may require perfect balance yet be made of two parts of unknown density bolted together. Here one must ask if there are enough "degrees of freedom" to satisfy the constraint. In this case, the unknown density ratio between the parts is no problem, so long as we are free to adjust the size of at least one part to bring the assembly into balance. This gives us the degree of freedom we need to compensate for the uncertainty.

This line of reasoning shows how the exploratory engineer can make a solid case for a device despite uncertainties about many of its properties and design details. It also shows how uncertain properties can create *compensating* uncertainties in design details (as in the example given, where an uncertain density leads to a corresponding uncertain size). This shows that "uncertainties" can come in planned sets that cancel out rather than adding up. The notion of uncertainty in exploratory engineering plays other tricks. Ignorance of these can lead one to confuse confidence-building factors with confidence-eroding factors. We need to avoid confusions about uncertainty, especially in large systems of ideas.

Uncertainties play different roles in science, standard engineering, and exploratory engineering. The usual intuitive rule about uncertainty in large sets of ideas or proposals is simple: If a conclusion or design rests on layer upon layer of shaky premises, it will surely fall. But this intuition sometimes misleads. To see where it works and where it doesn't, consider an imaginary proposal—a theory—in science, and another—a design—in engineering. Each proposal will have five essential parts and ten equally plausible possibilities for each part.

The theory might have to explain: (1) what something is; (2) where it came from; (3) how it survived the last million years; (4) why it hasn't been detected with X-rays; and (5) what it does when baked. Only one possibility can be right for each part of the theory, so with our assumed ten equally plausible possibilities for each part, a choice will have only a one-in-ten chance of being right. For a specific version of the theory, the chances of getting all five parts right (assuming no additional data) are no better than one-tenth to the fifth power. For a theory to be true, all its parts must be true, and so for any specific version the odds against it are at least one hundred thousand to one.

This artificial example shows how uncertainties combine in building real scientific theories: They combine adversely indeed. A scientific theory is a single-stranded chain that can break at any link. A chain with many dubious links is almost certainly worthless. This shapes the scientist's attitude toward uncertainty.

Uncertainties in exploratory engineering work in a different way. Consider a superficially similar design problem: a mechanism requiring five essential parts, with (again) ten equally plausible possibilities for each part. The design might require: (1) a power supply; (2) a motor; (3) a speed controller; (4) a locking device; and (5) an output shaft. But here, more than one possibility may work: Unlike theoretical proposals, engineering possibilities are not mutually exclusive. What is more, in exploratory engineering the typical problem is to build a case for the workability of the mechanism, not to specify a detailed, workable design—it is enough for there to be *one* working possibility.

In accord with these points, imagine that each of the ten possibilities for a part has a 50-percent chance of working. The chance of all ten possibilities failing, leaving no workable design for the part, is 0.5 to the tenth power—less than 1 in 1,000. With five parts facing this risk of unworkability, the overall chance that some essential part won't be possible is less than 5 in 1,000, making the overall probability of a workable design better than 99.5 percent. Real examples can give even better results: There may be a hundred ways to build each part, and several may be essentially sure bets.

In exploratory engineering, the "uncertainty" resulting from many possibilities may lead to a virtual certainty that at least one will work. An exploratory design concept can be like a massive, braided cable, in which many strands must fail before the link is severed. Uncertainties do not combine adversely, as do the superficially similar uncertainties of science (in science, a closer parallel would be a claim that *some* correct theory can be found, but even this suffers from the problem that only one choice can be right, and that choice may not be known). Standard engineering, however, is a bit closer to science in this regard.

In an idealized competitive world, only the best design would do. And choosing the *best* design, like choosing the true theory, means choosing the uniquely right possibility for each part. In the real world, the best isn't necessary, but competitive pressures still narrow the acceptable choices.

Further, in standard engineering it isn't enough to establish that there is a workable design somewhere in a forest of alternatives—one must propose a specific design, build it, and live with the consequences. Time and budgets are limited, and the failure of a large system may leave no resources for another try. In our model above, this would mean making five choices with a 50-percent chance of success with each part, making the overall chance of success about 3 percent. (This obviously motivates careful testing of parts before building systems.)

All these factors combine to make exploratory engineering more feasible than it might seem. Designs need not be competitive with other, similar designs; they need only be workable. To make them workable, they can be grossly overdesigned to compensate for uncertainties. Since their purpose is to provide a case for a possibility, not a blueprint to guide manufacture, they can omit details and include room for adjustment. All this aids in building a solid case for specific kinds of mechanisms, and concepts for whole systems of mechanisms, and concepts for whole systems of mechanisms can be solid even if they are built on layer upon layer of shaky cases for specific parts.

═══

In the first half of the twentieth century, work in exploratory engineering persuaded knowledgeable individuals that space flight would be possible. Today, the most important field for exploratory engineering is perhaps nanotechnology: It is clearly foreseeable and will be of immense practical importance. It can serve as a prime specimen of the process.

Nanotechnology is (or, rather, *will be*) a technology based on a general ability to build objects to complex, atomic specifications. We live in a world made out of atoms, and how those atoms are arranged makes a tremendous difference. This is why nanotechnology will make a tremendous difference.

Today's technology is a *bulk technology:* It handles atoms not as individuals, but as crowds. We make things by heating, stirring, molding, whittling, spraying—all processes that move trillions of atoms at a time, with only crude control over the patterns they form. Chemistry and genetic engineering use clever bulk-technology techniques to achieve specific molecular results, but only within narrow limits. No technology today gives us a general ability to build objects atom by atom.

Nanotechnology will be based on molecular machines and molecular electronic devices. With computers and robotic arms smaller than a living cell, it will enable the construction of almost anything, building up structures atom by atom. Among the products will be:

- Pocket computers with more memory and computational capacity than all the computers in the world today put together
- Large objects, such as spacecraft, made from light, superstrong materials and as cheap (pound for pound) as wood or hay
- Machines able to enter and repair living cells, giving medicine surgical control at the molecular level

How can one draw such conclusions? A more detailed exposition is spread over several papers and a book (*Engines of Creation:* Anchor/Doubleday, 1986), but the outlines are straightforward.

═══

The idea of nanotechnology resulted from applying an engineering perspective to the discoveries of molecular biology, and one path to nanotechnology lies through further advances in biotechnology. Regardless of how nanotechnology emerges, however, the facts of molecular biology provide a direct demonstration of principles that can be used by future molecular machines. (A rule of exploratory engineering: If one knows that it happens, one can assume it is possible.)

Cells contain enzymes, some of which function as molecular jigs and machine tools, assembling small molecules to build larger molecules. Enzymes themselves are made by ribosomes (which are in turn made by enzymes and earlier ribosomes). Ribosomes are complex molecular machines, programmed by the genetic system. Together, enzymes and ribosomes demonstrate that molecular reactions can be guided by molecular machines under programmed control, building up complex structures—including more molecular machines.

One path to nanotechnology, then, would use these biochemical machines to build new machines of our own design. This, however, means learning how to design protein molecules (the products of ribosomes), and this is a tough job. The biotechnology industry is hard at work on this problem, and making slow but real progress. Another path to nanotechnology would use chemical techniques to build nonprotein molecular machines—using nature's principles, but not nature's materials. The 1987 Nobel Prize in chemistry was presented to Jean-Marie Lehn, Donald Cram, and Charles Pedersen for work that leads in this direction. Yet another path would use the technology of the scanning tunneling microscope—which can position a needle with atomic precision near a surface—to manipulate molecules and so build molecular machines. No one has yet built a specific molecular structure this way, but the technology is still young, having been announced in 1982.

Thus, multiple paths lead from present technology toward a technology able to build complex molecular structures, including molecular machines able to build better molecular machines. The conclusion that we can build such machines gains strength from what they may be termed our "uncertainty regarding how to proceed"—that is, from the presence of many apparently workable options. This uncertainty does not spill over into nanotechnology itself, however, because all these developmental paths lead to the same destination.

Full-fledged nanotechnology will rely on molecular machines able to position reactive molecules to atomic precision, building up complex structures a few atoms at a time. These molecular construction machines are called "assemblers"; some advanced versions will resemble submicroscopic industrial robots. Nanotechnology is virtually synonymous with assembler-based technology, since only assemblers seem likely to give us the control needed to build complex structures to atomic specifications.

Nanotechnology is not defined by its size, even though the prefix "nano" means "billionth," just as "micro" means "millionth." Not everything that produces micrometer-scale objects qualifies as microtechnology: Exhaling particles of cigarette smoke fails the test. Likewise, not everything that produces nanometer-scale objects qualifies as nanotechnology: Burning hydrogen to make water molecules and etching ultrafine lines on a silicon chip fail the test. Nanotechnology, like microtechnology, is characterized not by the scale of its products but by the sorts of techniques used and the sorts of things those techniques can build. Microtechnology makes intricate micron-scale patterns using etching, vapor deposition, photolithography, and the like; nanotechnology will make intricate nanometer-scale patterns by building objects atom by atom.

Multiple paths lead to the land of nanotechnology. What lies within?

═══

To explore the domain of nanotechnology means exploring the world of things—especially molecular machines—that can be built using atoms as individually arranged building blocks. Knowledge of the forces within and between molecules tells us of the forces within and between the parts of molecular machines. The field of "molecular mechanics," developed by chemists, describes these forces and the resulting molecular motions, often quite well. The exploratory engineer can compensate for inaccuracies in modern molecular mechanics descriptions by overdesigning parts, by allowing large margins of safety, and by paying attention to the number of degrees of freedom in a design.

The most fundamental fact about molecular mechanics is that molecules can be thought of as objects. They have size, shape, mass, strength, stiffness, and smooth, soft, slippery surfaces. Large objects are made of many atoms; molecules are objects made of only a few atoms.

Molecular mechanics describes what happens when the atomic

bumps of one surface slide over the atomic bumps of another—and shows, surprisingly, that the resulting motion can be so smooth as to be almost frictionless at low speeds. This makes possible good bearings. Molecular mechanics can also describe how friction builds up with speed, but the analysis is complex and has not yet been done. Until it is, the exploratory engineer can design using low sliding speeds, and only then assume low sliding friction.

The story continues through other molecular devices. Meshing atomic bumps can serve as gear teeth. Helical rows of bumps can slide smoothly over other helical rows, serving as threads on nuts and screws. Tightly bonded sets of atoms—like tiny bits of ceramic, diamond, or engineering plastics—can form strong, rigid parts. Rotors, bearings, and electrodes can form electrostatic motors a few tens of billionths of a meter in diameter, which can produce an incredible amount of power for their size (many trillions of watts per cubic meter).

Motors, shafts, gears, bearings, and miscellaneous moving parts built in this way can combine to form robot arms less than one-tenth of a millionth of a meter long. Owing to a fundamental law relating size to rate of motion in mechanical systems, a robot arm this size can perform operations in one ten-millionth of the time required for an analogous device a meter long. Equipped with suitable tools, these arms can work as assemblers, building other machines at a rate of millions of molecular operations per second.

This sketches some of what seems clear from the exploration of nanotechnology. Many uncertainties remain, particularly in molecular electronics. Molecular *mechanical* devices can be analyzed using molecular mechanics and the familiar, Newtonian laws of motion (augmented by statistical mechanics, to describe thermal vibrations). Molecular *electronic* devices, in contrast, demand a quantum mechanical analysis, which is far more complex. Until a useful set of devices is designed and subjected to a clear, sound analysis, the exploratory engineer cannot design systems that assume the use of molecular electronics.

This might seem a grave limitation, since computer control has become so important in conventional engineering. Nonetheless, molecular *mechanical* computers (with properties supporting the above projection of "pocket computers with more memory and computational capacity than all the computers in the world today put together") can readily be designed. Although molecular electronic devices should be orders of magnitude faster, and may even outcompete molecular mechanical computers in all respects, the analyzability of molecular machines gives the mechanical approach a decisive advantage—not to the standard engineer of the future, but to the exploratory engineer of today. Analysis shows that molecular mechanical computers can be made to tolerate the jostling of thermal vibrations, and that these computers can run at about a billion cycles per second (somewhat faster than today's electronic machines), while consuming (roughly) tens of billionths of a watt of power. This technology will pack the memory and computational ability of a mainframe computer into the volume of a bacterial cell.

These mechanical nanocomputers can control assembler arms, directing their work. If a computer contains instructions (on molecular tape, say) for constructing a copy of an assembler and its raw-materials feed system, a copy of the computer, and a copy of a tape-duplicating machine, then the whole system can build a copy of itself. In short, it could replicate like a bacterium. Calculations indicate that a replicating assembler system of this sort could copy itself in less than an hour (remember the speed of assembler arms), using cheap industrial chemicals as raw materials. It is left as an exercise for the reader to calculate how long it would take one replicator, with a mass of (say) one-trillionth of a gram, to convert a million tons of raw materials into a million tons of replicators. (Hint: A ton is a million grams, and the answer is measured in days.) Slower-working replicators could run on air, water, sunlight, and a pinch of materials. For safety they would need reliable, built-in limits to growth, but that is an issue addressed elsewhere.)

These results, plus the observation that molecular machines can build large things, such as redwood trees, lead to the conclusion that teams of replicating assemblers can build large objects. By building atom by atom, they can build these objects from materials that are today impractical for structural engineering—materials such as diamond. This supports the above projection that nanotechnology will enable the construction of "large objects, such as spacecraft, made from light, superstrong materials and as cheap (pound for pound) as wood or hay." This, in turn, will make possible inexpensive housing, consumer goods, spacecraft, and so forth. These can be as inexpensive as other, more complex products of self-reproducing, solar-powered molecular machines: Crabgrass is harder to synthesize than diamond, unless one has crabgrass seeds to help. With seeds, making crabgrass is no trouble at all; with suitably programmed replicators, making simple things like spacecraft will be equally convenient.

The last projection listed above, "machines able to enter and repair living cells, giving medicine surgical control at the molecular level," is in a special category of difficulty and importance. The essential argument is simple. We observe molecular machines working within cells and able to build anything found in a cell—this is, after all, how cells replicate themselves. We observe that molecular machines can enter tissues (as white blood cells show), enter cells (as viruses show), and move around within cells (as molecular machines inside cells show). Molecular machines can also recognize other molecules (as antibodies do) and take them apart (as digestive enzymes do). Now combine these abilities (to enter tissues and cells and recognize, tear down, and build molecular structures) with control by nanocomputers, and the result is a package able to enter and repair living cells—almost.

The only substantial reservations about this conclusion involve knowledge and software. Knowledge about cells, and the difference between diseased and healthy cells, is not the problem. Though this will be new scientific knowledge, and hence not predictable in its particulars, acquiring that knowledge is a problem amenable to an engineering solution. In the early 1960s one could project that we would learn the particulars of cell structure through nanotechnology (if not sooner, by other means). Knowledge of how to build software able to perform complex cell diagnosis and repair processes, however, is harder to project. This will involve building software systems of greater complexity than those managed in the past, requiring new techniques. Progress in these aspects of computer science has been swift but is hard to project.

Simple cell repair systems are within the range of confident projection today. Repair systems able to tackle more complex problems (such as repairing severe, long-term, whole-body frostbite) seem likely, and can be analyzed in many details today, but discussing them appears to involve an element of speculation regarding future progress in software engineering.

━━━

As the last example shows, on the frontier of the domain of exploratory engineering lie problems characterized more by their complexity than by their physical novelty. These are problems whose solutions will demand new design techniques.

Attempts to project new design techniques can run afoul of a problem like that of trying to project future scientific knowledge: New design techniques will often stem from new insights, and if we could say what the insights will be, we would have already had them.

Other attempts pose fewer problems. For example, if faster, cheaper computers are the key to a new design technique, then the possibility of that technique becomes fair game for exploratory engineering. Curiously, the idea of building devices that can think like engineers, but far faster, can be examined in this way. While this capability seems most likely to be achieved in some novel way based on new insights, it *could* be achieved through the use of nanotechnology to study and model, component by component, the functions of the brain. Then, without necessarily *understanding* how the brain works, one could build a fast, brainlike device. This would not really be artificial intelligence, however; it would merely provide a new physical embodiment for the intricate patterns of naturally evolved intelligence.

Issues of complexity are not central to nanotechnology itself. Assemblers need be no more complex than industrial robots. Nanocomputers need be no more complex than conventional computers. Even replicating assembler systems seem no more complex than modern automated factories. The fundamental capabilities of nanotechnology thus entail no more complexity than we have already mastered. Though nanotechnology will permit engineers to build systems of unprecedented complexity in a tiny volume, it does not *demand* that they do so.

◻▭◻

Exploratory engineering has limits, but there is much to be achieved within those limits. Successful exploratory engineering can be of great value from a variety of perspectives. For the technophile, it can reveal directions for research that promise great benefit, increasing the returns on society's investment. For the technophobe, it can reveal some of the dangers for which we must prepare, helping us handle new abilities with greater safety. Success in exploratory engineering, and in heeding its results, may be a matter of life and death. If so, then it seems we should make some effort to become good at it.

The first step is to recognize and criticize it. No field can flourish unless it is recognized as having intellectual standards to uphold; to have a discipline, one must have discipline. Exploratory engineering has too often been seen as not-science and not-(standard) engineering, and hence lumped together with scientific speculation and science fiction. Since speculation and fiction make no pretense of solidity, they are not subject to rigorous criticism to separate the solid from the erroneous. Exploratory engineering is different, and should be criticized on the basis of its aims. Those aims, again, are not to prophesy new scientific knowledge, not to prognosticate the details of the competitive designs of the future, but to make a solid case for the feasibility of certain classes of future technology.

In the absence of criticism, nonsense flourishes. Where nonsense flourishes, sense is obscured. We need to recognize and criticize work in exploratory engineering, in order to make a bit more sense of our future.

Counting the Ways:
The Scientific Measurement of Love

ROBERT STERNBERG

Each aspect of love can be viewed as generating a different side of a triangle. Each individual in a given loving relationship has his or her own triangle, and these triangles may or may not correspond closely. Moreover, each partner has an ideal triangle, which represents the partner they would like to have, and an action triangle, which represents the extent to which their actions speak for their feelings in each of the three domains of intimacy, passion, and commitment.

SHE: Do you love me?

HE: Of course I do.

SHE: How much do you love me?

HE: Oh, lots and lots.

SHE: But really, how much?

HE: Eight, on a one-to-nine scale.

In a romantic conversation, his last remark might seem somewhat incongruous, but in the scientific study of love, it would not be incongruous at all: The claim of this article is that love, like other psychological constructs, can be studied scientifically, and that, like other psychological constructs, it can also be quantified and measured.

There are good reasons to study love scientifically, despite the resistance psychologists studying love scientifically often encounter from laypeople and scientists alike. A first and not inconsequential reason is simply to show it can be done: Senator Proxmire, in bestowing a Golden Fleece Award upon an internationally famous love researcher, was not alone in his belief that the study of love should be left to poets. Even many psychologists have been skeptical as to whether love can be studied scientifically, and as a result, the field languished for many years. Love is a key aspect of many interpersonal relationships, and it is in some ways rather odd that, for so many years, the study of interpersonal relationships flourished in psychology at the same time that the study of love languished.

A second reason is that with the divorce rate in the United States approaching 50 percent, there is a societal as well as a scientific need to understand what love is, and what leads it either to grow or to die. If we could measure love, then we might be able to ascertain what kinds of feelings and actions lead to its growth or decline. We might also be able to use these measurements to predict which relationships are likely to succeed and which to fail.

Third, if we could measure love in close relationships, we ultimately might even be able to use these measurements to intervene therapeutically and suggest to individuals and couples ways in which they could improve their relationships. Rather than making guesses about which aspects of relationships need improvement, we might be able to identify these aspects scientifically, and then set about correcting them, wherever possible.

We might worry that measuring love would simply tell us what Grandmother already knows—in quantified form with impressive statistics that Grandmother never needed. But the measurement of love has led to a number of scientific observations that are anything but routine. For example, Susan Grajek, a former graduate student at Yale, and I found that, on the average, whereas men love their lovers more than their best friend of the same sex, women love their lover and best friend of the same sex about equally, and actually like their best friend of the same sex somewhat more than their lover.[1] Michael Barnes, a graduate student at Yale, and I found that the best predicter of satisfaction in an intimate heterosexual relationship for a given partner is not how much one loves one's partner, or even how much one's partner loves oneself, but, rather, the difference between how much one perceives one's partner to love oneself and how much one ideally wants to be loved by one's partner.[2] We can feel "starved" by too little love from our partner, but we can also feel "suffocated," in some relationships, by too much of it. In predicting satisfaction, Michael Barnes and I found that there is not a particularly high correspondence between how much one thinks one's partner loves one and how much one's partner actually loves one. Moreover, in predicting satisfaction, once one has entered into the prediction equation how much one thinks one's partner loves one, how much one's partner *actually* loves one makes no difference to one's satisfaction. In short, the findings that emerge from scientific studies of love are by no means routine or obvious and do, indeed, take us beyond Grandmother's knowledge of love and its consequences.

The study of love has by no means been the exclusive province of psychologists. Aristophanes, in Plato's *Symposium,* suggests that people were originally of three kinds—male, female, and androgynous. Androgynous individuals were round, with their backs and sides forming a circle. They had one head with two faces looking in opposite directions, and had four hands and feet, as well as two sexual organs. But the gods came into conflict with humanity, and in their determination to cripple humanity, the gods cut all humanity into two parts. The males became homosexual; the women, lesbian; and the androgynous, heterosexual: Each of the three kinds, to this day, is seeking his or her lost part, seeking to complete himself or herself.

Many philosophers and writers since Plato have had a great deal to say about love, but in psychology, the origins of theorizing about love, as about so many other things, date most clearly to Sigmund Freud. Freud, like many others, believed that the origins of love are sexual: The infant is narcissistic, loving only himself or herself.[3] Soon, the love of the infant is directed toward the parents, and especially the parent of the opposite sex; but later, because of incest taboos, this love is ultimately directed toward an opposite-sex partner outside the immediate family. For Freud, love is essentially sublimated sexuality—that is, sexuality redirected in a socially appropriate way.[4] Harry Harlow, formerly a psychologist at the University of Wisconsin, viewed love in terms of attachment toward another, and Erich Fromm identified love with feelings of care, responsibility, respect, and knowledge of another.[5] Abraham Maslow, formerly a psychologist at Brandeis University, distinguished between two kinds of love: deficiency love, which arises out of a person's insecurities and lower-level emotional needs; and being love, which arises out of a person's higher-level emotional needs, and especially the desire for self-actualization and the actualization of another.[6] Maslow's deficiency love is similar to Theodor Reik's conception of love as the search for salvation—finding in another what one cannot find in oneself.[7] Ultimately, this conception probably harks back to Plato's ideas about the search for completion.

The scientific study of love was placed on an entirely new footing through the research of a graduate student in psychology at the University of Michigan, Zick Rubin (now at Brandeis). Rubin believed that love could be quantified and, hence, measured.[8] In a landmark article on "the measurement of romantic love," he proposed two scales, one of which was purported to measure love, and the other of which was purported to measure liking.[9] Subjects were asked to indicate their level of agreement with each of a series of statements, and their levels of agreement were quantified on a one-to-nine scale. The love scale had items measuring what Rubin believed were three main aspects of love: attachment, caring, and intimacy. Examples of scale items measuring each of these characteristics of love were: "It would be hard for me to get along without —————" (attachment); "I would do almost anything for —————" (caring); and "I feel that I can confide in ————— about virtually everything" (intimacy). Three characteristics of liking were measured by the liking scale—admiration, respect, and perceived similarity. Examples of liking-scale items were "I feel that ————— is unusually well adjusted" (admiration); "I have great confidence in —————'s good judgment" (respect); and "I feel that ————— and I are quite similar to one another" (perceived similarity). The full love and liking scales, containing thirteen items each, can be found in Rubin's *Liking and Loving: An Invitation to Social Psychology.*[10]

Of course, anyone can claim that a proposed scale measures love or liking. For any scale to win acceptance, it has to be shown to be reasonably internally consistent, or reliable, and to have some kind of empirical validity. Rubin, aware of this problem, showed both the reliability and validity of his scales. Not only were the scales internally consistent, but they proved to have a number of interesting correlates in studies of college students. For example, love-scale scores were fairly highly correlated ($r = 0.59$ for both men and women, on a 0 [low]–to–1 [high] scale) with partners' estimates of the probability that they would marry, whereas liking scores were only moderately correlated with these estimates ($r = 0.35$ for men and 0.32 for women). Love scores were also correlated with measured levels of intimate disclosure to one's partner ($r = 0.46$ for men and 0.51 for women), with liking scores also showing positive but lower correlations (r = 0.21 for men and 0.37 for women). Love scores were also correlated with the amount of gazing into each other's eyes partners did while waiting to participate in the experiment; and for those who believed love to be an important aspect of deciding upon a marital partner, love scores were correlated with the longevity of the intimate relationship. Loving and liking are themselves correlated, although in Rubin's studies, the correlations are only in the middle ranges (r = 0.56 for men and 0.36 for women).

In sum, Rubin's work essentially started a new field—the measurement of love. Investigators took off from Rubin's seminal work in a number of directions, most of which attempted to put the study of love on a more theoretical footing. Rubin's theory, although based to some extent on factor-analytic data, was largely intuitive. Some of the subsequent work has attempted to elaborate on and modify our theoretical understanding of love.

Some investigators have explored ways of measuring particular aspects of love. For example, Elaine Hatfield, a psychologist at the University of Hawaii, has concentrated on the measurement of passionate love, whereas Ellen Berscheid, a psychologist at the University of Minnesota, has concentrated on emotion and intimacy in close relationships. [11] Other investigators have attempted to measure both liking and loving, conceptualizing them in ways somewhat different from Rubin. Keith Davis, a psychologist at the University of South Carolina, for example, has proposed that love is liking plus physical attraction and caring beyond that found in liking. [12] But still other investigators have tried to generate measurements of love based on global theories of the nature of love.

One attempt to place the study of love on a more theoretical footing was made by an investigator whose earlier research had been in a field seemingly as distant from the field of love as could be possible. The distant field was intelligence; the investigator was myself. In an initial article, Susan Grajek and I made what might seem like a strange proposal—that three alternative structural models of intelligence might be carried over to the study of love. [13] These three models are shown in the figure on page 100.

The first model was based upon the thinking of Charles Spearman, a British psychologist at the turn of the century who had

"Spearmanian" Model

"Thomsonian" Model

"Thurstonian" Model

Three alternative models of love.

proposed a "two-factor" theory of intelligence, according to which there is a general factor pervading all intellectual performances, as well as a set of less interesting specific factors, with each of the latter limited to performance on a single task.[14] Similarly, one might conceptualize love in terms of a single general factor, which would be an undifferentiated "glob" of highly positive and emotion-charged affect that is essentially nondecomposable. To experience love would be to experience this glob of highly positive affect.

The second model was based upon the work of Godfrey Thomson, a British psychologist, who had proposed that the mind possesses an enormous number of bonds, including reflexes, habits, and learned associations.[15] Intellectual performance on any one task would activate a large number of these bonds. Related tasks, such as those used in intelligence tests, would sample overlapping subsets of bonds. A factor analysis of a set of tests might therefore give the appearance of a general factor, when, in fact, what is common to the tests is a multitude of bonds. In terms of a structural model of love, one might conceptualize love in terms of a set of affects, cognitions, and motivations that, when sampled together, yield the composite emotion that we label love. In this view, though, the composite is not as undifferentiated unity; rather, it can be decomposed into a large number of underlying bonds that tend to co-occur in certain close relationships and that in combination result in the global feeling that we view as love.

The third model was based upon the work of Louis Thurstone, a psychologist of the early twentieth century at the University of Chicago who had proposed a theory of intelligence comprising seven primary and equally important factors, such as verbal comprehension, memory, and inductive reasoning.[16] The underlying idea was that intelligence is composed of a relatively small set of correlated primary mental abilities. Applying this notion to love, one would emerge with a theory viewing love in terms of a small, consistent set of emotions, cognitions, and motivations that are of approximately equal importance and salience in the overall feeling we describe as love. Love is not one main thing, whether decomposable (Thomsonian model) or not (Spearmanian model). Rather, it is a set of primary structures that are best understood separately rather than as an integrated whole.

Susan Grajek and I administered the Rubin Love and Liking Scales as well as using an instrument based on the Levinger, Rands, and Talaber Interpersonal Involvement Scale. Our subjects were eighty-five adults from the greater New Haven area who had answered a newspaper advertisement.[17] All were primarily heterosexual, eighteen years of age or older, and had been involved in at least one intimate relationship. They filled in the Rubin and Levinger *et al.* scales not only for their lover, but for their mother, father, sibling closest in age, closest same-sex friend, and oldest child. (Data about children were discarded because the number of subjects having children was not large enough for adequate statistical analysis.) Two main statistical techniques were then applied to the data:

factor analysis, which seeks to discover the latent structure underlying a data set in terms of continuous dimensions; and cluster
analysis, which seeks to discover latent structure in terms of discrete
groupings, or clusters.

The factor-analytic results for all of the interpersonal relationships
indicated a strong general factor. The general factor was labeled
"interpersonal communication, sharing, and support," based upon
items from the Rubin and Levinger *et al.* scales that showed particularly high correlations with the factor. The emergence of this general factor supported the models of Spearman and Thomson, but not
of Thurstone. In order to distinguish between the former two models, it was necessary to discover whether the general factor could be
broken down. Analysis revealed ten distinct and identifiable clusters, thereby supporting the model of Thomson over that of Spearman. The ten clusters were a desire to promote the welfare of the
loved one, experienced happiness with the loved one, a high regard
for the loved one, being able to count on the loved one in times of
need, a mutual understanding with the loved one, sharing of oneself
and one's possessions with the loved one, receipt of emotional support from the loved one, giving of emotional support to the loved
one, intimate communication with the loved one, and valuing the
loved one in one's life.

As sometimes happens after findings on love are reported in a
scholarly scientific journal, the general media picked up on some of
our results. The reporters who talked to me all seemed to make the
same point: Isn't the view of love that emerged from this study a
rather tame one, one that does not seem to take into account the
hotter, or more passionate, aspects of love? At first, I argued that
these hotter aspects of love are concomitants of love rather than part
of the love itself. This line wore rather thin after a while, though,
and I began to think about how the Sternberg-Grajek "Thomsonian" model might be elaborated to view love in a broader way. In
particular, I became concerned that the Rubin and Levinger *et al.*
scales might have sampled only a narrow subset of love as a complete
entity, and that the results that had emerged from our study were all
thereby conditionalized upon the possibly narrow sampling of aspects of love in the questionnaires.

According to my triangular theory of love, love comprises three components: intimacy, passion, and decision/commitment. Each component manifests a different aspect of love.[18]

Intimacy refers to feelings of closeness, connectedness, and bondedness in loving relationships. It thus includes within its purview those feelings that give rise, essentially, to the experience of warmth in a loving relationship. The ten clusters obtained in my study with Susan Grajek mentioned earlier represent some of the elements of intimacy: desire to promote the welfare of the loved one, being able to count on the loved one in times of need, and mutual understanding with the loved one.

Passion refers to the drives that lead to romance, physical attraction, sexual consummation, and related phenomena in loving relationships. The passion component includes within its purview those sources of motivational and other forms of arousal that lead to the experience of passion in a loving relationship. It includes what Elaine Hatfield and William Walster refer to as "a state of intense longing *for union* with the other."[19] In a loving relationship, sexual needs may well predominate in this experience. However, other needs—such as those for self-esteem, nurturance, affiliation, dominance, and submission—may also contribute to the experiencing of passion.

Decision/commitment refers, in the short term, to the decision that one loves a certain other, and in the long term, to one's commitment to that love. These two aspects of the decision/commitment component do not necessarily go together, in that one can decide to love someone without being committed to the love in the long term, or one can be committed to a relationship without acknowledging that one loves the other person in that relationship.

Each aspect of love can be viewed as generating a different side of a triangle. Each individual in a given loving relationship has his or her own triangle, and these triangles may or may not correspond closely. Moreover, each partner also has an ideal triangle, which represents the partner they would like to have, and an action triangle, which represents the extent to which their actions speak for their feelings in each of the three domains of intimacy, passion, and commitment. Geometrically, the size of a triangle represents the amount of love one partner experiences toward the other, and the shape of a triangle represents the balance of the three components of love in the relationship, as shown in the figure below.

The three components of love generate eight possible subsets when considered in combination. Each of these subsets gives rise to a different kind of love: nonlove, liking, infatuated love, empty love, romantic love, companionate love, fatuous love, and consummate love. The particular combinations of components generating each of these subsets are shown in the table on page 105.

Nonlove refers simply to the absence of all three components of love. Liking results when one feels closeness, bondedness, and warmth toward another without feeling intense passion or the desire

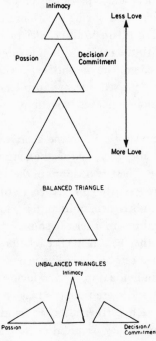

Different amounts and balances of love according to the triangular theory of love.

Taxonomy of Kinds of Love

		Component		
		Intimacy	Passion	Decision/ Commitment
1.	Nonlove	−	−	−
2.	Liking	+	−	−
3.	Infatuated Love	−	+	−
4.	Empty Love	−	−	+
5.	Romantic Love	+	+	−
6.	Companionate Love	+	−	+
7.	Fatuous Love	−	+	+
8.	Consummate Love	+	+	+

NOTES: These kinds of love represent limiting cases based upon the triangular theory. Most loving relationships will fit between categories, because the various components of love are expressed along continua, not discretely. " + " = component present; " − " = component absent.

for long-term commitment. Infatuated love is essentially passionate "love at first sight" and results from high psychophysiological arousal in the absence of intimacy or commitment. Empty love results when one feels committed to a "loving" relationship in the absence of feeling either intimacy or passion. Romantic love results from a combination of intimacy and passion, such that one feels romantic toward another but friendly as well. Companionate love emanates from a combination of intimacy and commitment and tends to characterize long-term relationships in which the passion that was once there has died. Fatuous love, deriving from passion and commitment in the absence of intimacy, results when a couple commit themselves on the basis of attraction without enabling or possibly allowing the development of intimacy. Consummate love results from the combination of intimacy, passion, and commitment and is the love toward which many heterosexuals appear to strive in close relationships.

Each of the proposed components of love shows a different time course, with both intimacy and passion exhibiting some suscepti- bility to habituation over time. In other words, over time, the components of love may decrease unless efforts are made to maintain variety and challenge in a relationship. Much of the success of a relationship can depend upon the couple's success in counteracting habituation.

I performed a validation of my theory of love in which I sought simultaneously to test the theory and a measure of it. The subjects in my study were eighty-four New Haven area adults, equally divided among men and women and ranging in age from nineteen to sixty-two, with a mean age of twenty-eight. All were involved in intimate heterosexual relationships whose duration ranged from about a month to twenty-two years with a mean of four and a half years.

All subjects received a series of questionnaires, including the Sternberg Triangular Love Scale,[20] as well as the Rubin scales, as described earlier. Included among the questionnaires was also a satisfaction measure that contained nine items querying the subjects regarding their satisfaction with their current intimate relationship. On my scale, subjects rated on a one (low)–to–nine (high) scale their responses to seventy-two statements, half of which reflected feelings and half of which reflected actions. The action questions were the same as the feeling questions, except that they were preceded by the phrase "My actions reflect . . ."

Twelve of the feelings questions were written to measure intimacy, twelve to measure passion, and twelve to measure decision/commitment. The different kinds of statements were intermixed in the questionnaire, so that subjects could not readily perceive which statements measured what. Of course, subjects were not informed in advance of the nature of the triangular theory or any other theory. Examples of intimacy, passion, and decision/commitment questions, respectively, are: "I strongly desire to promote the well-being of ————"; "Just seeing ———— is exciting for me"; and "I am committed to maintaining my relationship with ————." Half the subjects were instructed to rate all the statements for six different love relationships (mother, father, sibling closest in age, lover/spouse, best friend of the same sex, and ideal lover/spouse) in terms of how *important* each statement was, in the subjects' minds, to each of the six relationships. The other half of the subjects were instructed to rate the statements on how *characteristic* each was in their own lives for each of the six relationships. Importance is a value judgment, characteristicness a judgment of the actual state of an existing relationship.

The three subscales proved to be internally consistent. The overall scale reliability was in the high 0.90s (where 0 indicates no reliability and 1 indicates perfect reliability). Feelings and action ratings were very highly correlated (generally in the 0.90s), indicating that people generally believed that their patterns of actions reflected their feelings. However, the reflection was imperfect, because the means of the action ratings were lower than those of the feelings ratings. In other words, people's actions reflect their feelings, but not the full extent of them. The three subscales were fairly highly intercorrelated with one another, although the extent of the correlations differed across relationships. Overall, the highest correlation for characteristicness ratings was between intimacy and commitment (0.81), and the lowest between passion and commitment (0.68—where 0 indicates no relationship and 1 indicates a perfect relationship).

Factor analyses were performed on both the characteristicness and importance ratings in order to determine whether the patterns of inter-item correlations accurately reflected the theory underlying the scale. Factor analysis is a statistical technique that identifies underlying sources of variations in a set of data. Three factors emerged from the overall characteristicness ratings. These factors reflected the constructs of decision/commitment, intimacy, and passion. A comparable factor analysis of the overall importance ratings resulted in four factors, which were the same as the factors for the characteristicness ratings except that the decision and commitment subcomponents of the decision/commitment component split off from each other.

Scores from both Rubin's scale and my scale were used to predict scores on the satisfaction questionnaire the subjects had received. For my subscales, correlations with the satisfaction questionnaire were 0.86, 0.77, and 0.75 for intimacy, passion, and decision/commitment, respectively, on the 0-to-1 scale. Comparable correlations for the Rubin scales were 0.36 and 0.59 for the liking and loving scales, respectively. These correlations indicate that the scale based upon my triangular theory was quite successful in predicting satisfaction with a current intimate relationship, and thus that the scale might be useful for practical as well as research purposes.

Although the results for my scale were quite good, they were by no means perfect. A couple of items just did not work. "I view my commitment to ———— as a matter of principle" proved to be a poor item, showing only trivial correlations with scores on the decision/ commitment scale of which it was a part (after total subscale scores were corrected for contribution of that item). Another item, "My relationship with ———— is very 'alive,' " was supposed to measure passion, but proved to correlate more highly with intimacy than with passion scores (again corrected for the contribution of the item to the subscale score).

An alternative approach to a theory of love has been taken by investigators studying styles of loving. John Lee, a sociologist at the University of Toronto, has proposed a set of different "styles" of loving.[21] According to Lee, there are three primary styles of love: eros, the love style characterized by the search for a beloved whose physical presentation of self embodies an image already held in the mind of the lover; ludus, which is Ovid's term for playful or game-like love; and storage, a style based on slowly developing affection and companionship. There are also three secondary styles: mania, a love style characterized by obsession, jealousy, a great emotional intensity; agape, which is altruistic love in which the lover views it as his or her duty to love without expectation of reciprocation; and pragma, a practical style involving conscious consideration of such characteristics of the loved one as social class, religion, and ethnic group.

These styles are not unrelated to those in my triangular theory. For example, eros might be regarded as fairly close to romantic love, and storage as fairly close to companionate love. But there are also some clear noncorrespondences. Ludus, for example, would not be regarded as a kind of love in my triangular theory, but rather as a style of interrelating that people can use in various kinds of loving relationships.

The first attempt to measure Lee's six love styles was made by Thomas Lasswell and Terry Hatkoff, clinical psychologists, who devised a fifty-item true/false scale to measure the six styles of loving.[22] A more extensive attempt to measure the constructs of the theory and then to test the theory has been made by Clyde Hendrick and Susan Hendrick of Texas Tech University.[23] In the series of two studies, the Hendricks used a forty-two–item questionnaire with seven items measuring each of the styles of love on two large samples. Styles of loving were shown to be related to various demographic and gender-related variables. For example, males were significantly more ludic than females, and Asians were less erotic in orientation than were blacks or Hispanics.

A quite different approach to styles of loving has been taken by Philip Shaver, currently at the State University of New York at Buffalo, and his colleagues.[24] These investigators have followed John Bowlby, a British scholar, in suggesting that styles of romantic love seem to correspond to styles of infants' attachment to their mothers.[25] Shaver and his colleagues have used Mary Ainsworth's attachment theory to elaborate upon this notion.[26]

According to these investigators, romantic lovers tend to have one of these different styles in their relationships. Their style is a stable characteristic and derives in part from the kind of attachment they had to their mother when they were young. Secure lovers find it relatively easy to get close to others. They also find they can be comfortable in depending on others, and in having others depend on them. They do not worry about being abandoned or about someone getting too close to them. Avoidant lovers are uncomfortable being close to others. They find it difficult to trust others completely and difficult to allow themselves to depend on others. They get nervous when anyone gets too close, and often they find that their partners in love want to become more intimate than they find comfortable. Anxious-ambivalent lovers find that others are reluctant to get as close as they would like. They often worry that their partners do not really love them or won't want to stay with them. They want to merge completely with another person, and this desire sometimes scares others away.

In their research, Shaver and his colleagues found that about 53 percent of their subjects were secure, 26 percent avoidant, and 20 percent anxious-ambivalent, percentages that correspond roughly to percentages of the three kinds of attachment relations in infants. Shaver and Cindy Hazan, of the University of Denver, have developed a scale for measuring tendencies toward each of the three kinds of love, and their results are consistent with the notion that people tend toward one or another of the attachment styles. Shaver's work is complementary to that of theorists such as myself and Lee: Different styles of attachment could lead to different preferred shapes of triangles for people or, in Lee's theory, to different styles of loving.

Since the 1970s, rapid progress has been made in the conceptualization and operationalization of love. Rubin's seminal work showed that it is possible to theorize about love in a way that is empirically testable, and subsequent developments have refined and expanded our notions both about love and how it can be measured. Although thinking about love goes back a long, long way, the scientific study of love is, by almost any standard, still in its infancy. Perhaps the greatest revelation of this work is not that any single theory or measure is clearly preferable—the field is too young for this kind of conclusion—but that love can be studied scientifically in the first place.[27] What had seemed to be as elusive and as ineffable as any psychological construct has proven itself tractable in terms of the same kinds of scientific methods used in other domains of psychology. Love may or may not conquer all, but it now appears that the study of love can be conquered by the theories and methods of science.

Currently, my graduate student Michael Barnes and I are pursuing an approach to the study of love that is quite different from the approaches described earlier. This approach concentrates on people's conceptions of love—what they think it is. We believe this direction to be a promising one, because what a person believes love to be will determine, in large part, how he or she relates to another in a loving relationship. We have found, for example, that people distinguish among three levels of intimacy—what is good for relationships in general, what is good for them individually, and what is good for them in a particular relationship with another. By studying both scientific and popular conceptions of love, it is possible to achieve a kind of balance in one's thinking about love that I believe could not be achieved by considering one kind of information alone.

NOTES

[1] Robert J. Sternberg and Susan Grajek, "The Nature of Love," *Journal of Personality and Social Psychology*, 1984, vol. 47, pp. 312–329.

[2] Robert J. Sternberg and Michael Barnes, "Real and Ideal Others in Romantic Relationships: Is Four a Crowd?" *Journal of Personality and Social Psychology*, 1985, vol. 49, p. 1589–1596.

[3] Sigmund Freud, *Three Contributions to the Theory of Sex* (New York: Dutton, 1905/1962).

[4] Sigmund Freud, "Certain Neurotic Mechanisms in Jealousy, Paranoia, and Homosexuality." In *Collected Papers* (London: Hogarth, 1922/1955), vol. 2, pp. 235, 240, 323.

[5] H. H. Harlow, "The Nature of Love," *American Psychologist*, 1958, vol. 13, pp. 673–685.

E. Fromm, *The Art of Loving* (New York: Harper, 1956).

[6] A. H. Maslow, *Toward a Psychology of Being* (Princeton, N.J.: Van Nostrand, 1962).

[7] T. Reik, *A Psychologist Looks at Love* (New York: Farrar & Rinehart, 1944).

[8] Zick Rubin, "Lovers and Other Strangers: The Development of Intimacy in Encounters and Relationships," *American Scientist*, 1974, vol. 62, pp. 182–190.

[9] Zick Rubin, "Measurement of Romantic Love," *Journal of Personality and Social Psychology*, 1970, vol. 16, pp. 265–273.

[10] Zick Rubin, *Liking and Loving: An Invitation to Social Psychology* (New York: Holt, Rinehart & Winston, 1973).

[11] Ellen Berscheid, "Emotion," in H. H. Kelley, E. Berscheid, A. Christensen, J. H. Harvey, T. L. Huston, G. Levinger, E. McClintock, L. A. Peplau, and D. R. Peterson, eds., *Close Relationships* (New York: Freeman, 1983), pp. 110–168.

Elaine Hatfield, "Passionate and Companionate Love," in R. J. Sternberg and M. Barnes, eds., *Psychology of Love* (New Haven: Yale University Press, 1988), pp. 191–217.

Elaine Hatfield and G. W. Walster, *A New Look at Love* (Reading, Mass.: Addison-Wesley, 1981).

[12] Keith Davis, "Near and Dear: Friendship and Love Compared," *Psychology Today*, February 1985, vol. 19, pp. 22–30.

[13] Robert J. Sternberg and Susan Grajek, "The Nature of Love," *Journal of Personality and Social Psychology*, 1984, vol. 47, pp. 312–329.

[14] C. Spearman, *The Abilities of Man* (New York: Macmillan, 1927).

[15] G. H. Thomson, *The Factorial Analysis of Human Ability* (London: University of London Press, 1939).

[16] L. L. Thurstone, *Primary Mental Abilities* (Chicago: University of Chicago Press, 1938).

[17] Robert J. Sternberg and Susan Grajek, "The Nature of Love," *Journal of Personality and Social Psychology*, 1984, vol. 47, pp. 312–329.
Zick Rubin, *Liking and Loving: An Invitation to Social Psychology* (New York: Holt, Rinehart & Winston, 1973).
George Levinger, M. Rands, and R. Talaber, *The Assessment of Involvement and Rewardingness in Close and Casual Pair Relationships,* National Science Foundation Technical Report DK (Amherst, Mass.: University of Massachusetts, 1977).

[18] Robert J. Sternberg, "A Triangular Theory of Love," *Psychological Review*, 1986, vol. 93, pp. 119–135.

[19] Elaine Hatfield and G. W. Walster, *A New Look at Love* (Reading, Mass.: Addison-Wesley, 1981).

[20] Robert J. Sternberg, *Construct Validation of a Triangular Theory of Love* (Manuscript submitted for publication, 1987).

[21] John A. Lee, *The Colors of Love: An Exploration of the Ways of Loving* (Don Mills, Ontario: New Press, 1973).

[22] T. E. Lasswell and M. E. Lasswell, "I Love You but I'm Not in Love with You," *Journal of Marriage and Family Counseling*, 1976, vol. 38, pp. 211–224.

[23] Clyde Hendrick and Susan Hendrick, "A Theory and Method of Love," *Journal of Personality and Social Psychology*, 1986, vol. 50, pp. 392–402.

[24] Philip Shaver and Cindy Hazen, "Romantic Love Conceptualized as an Attachment Process" (paper presented at the International Academy of Sex Researchers, Seattle, Wash., September, 1985).

Philip Shaver, Cindy Hazan, and Donna Bradshaw, "Love as Attachment: The Integration of Three Behavioral Systems," in R. J. Sternberg and M. Barnes, eds., *Psychology of Love,* pp. 68–99. (New Haven: Yale University Press, 1988).

[25] J. Bowlby, *Attachment and Loss—vol. 1, Attachment* (New York: Basic Books, 1969).

[26] M. D. S. Ainsworth, "The Development of Infant-Mother Attachment," in B. M. Caldwell and H. N. Ricciuti, eds., *Review of Child Development Research* (Chicago: University of Chicago Press, 1973), vol. 3, pp. 1–94.

[27] R. J. Sternberg, *The Triangle of Love: Intimacy, Passion, Commitment* (New York: Basic Books, 1988).

Bacterial Writing and the Metaphysics of Sex

DORION SAGAN

Is sex then part of a broader connecting phenomenon, a sort of parasexuality that brings together not only organisms of the same species but those of different species, as well as objects, molecules, friends, and ideas?

In two words, im possible.

—Goldwyn

The pen is mightier than the sword.

—Bulwer-Lytton

I

It seems impossible to talk, let alone write, objectively about sex because of the root metaphors of metaphysics—the basic, all-too-often unexamined prejudices underlying our thought, language, and discourse. We might call these governing metaphors metaforms because the influence they exert is quite literal. Becoming aware of them does not necessarily enable us to get around them. Metaphysical assumptions—including the assumption that metaphysics is philosophical nonsense—are all the more seductive when one is under their spell. Scientists, for example, often profess that metaphysics is of only passing interest to serious thought—for example as a phase, along with religion and magic, beyond which science has progressed. One should, therefore, begin a discussion of the scientific reality of sex by broaching the topic of sex's unexamined metaphysics, its underlying preconceptions, our inherited biases. Is it possible to be objective about sex? Obviously, when we treat each other as sex objects we are not being objective (despite the different meanings of this same word *object*). People are not simple objects of pleasure or specimens for scientific experiments but complex and mysterious beings. Using also the word metaphysics in a different sense, we might note that, according to some mystics, during the height of sexual intercourse, subject/object distinctions can break down. Then the lovers melt into each other; they retreat to an amniotic slipperiness; they come, ecstatically, out of themselves. An oceanic feeling of oneness, in which the act of coupling stands in symbolic relation to what would be a total mystical union, occurs. They may even realize all is one as they "die" into each other's arms.

Jacques Derrida, surely the most seminal philosopher of the twentieth century, has shown that what he calls the "metaphysics of presence" pervades Western thought. The metaphysics of presence lets us think in terms of stable constants, identities, determined references, and origins. But because all our language is built upon the root suppositions of Western metaphysics—or, what is nearly the same thing, all our underlying metaphysical suppositions are buttressed by the nature and flow of language—it would be impossible to state *definitively* what is meant by this expression. In general, however, the idea that there is an *arche,* a privileged cause or explanatory point of origin, to which things and events can be definitively traced, is put into question. In traditional archaeology, the bottom layers are assumed to be more primitive, "giving rise" to the upper ones; in a traditional hierarchy, the head is associated with the front, the top, the upper parts of the ruling principle or prince, the sun. Georges Bataille, in a perverse essay entitled *The Solar Anus,* pokes serious fun at the search for *arches* by attributing the thousands of lovers mating at any given moment on planet earth to the rolling wheels and driving pistons of steam engines coursing through landscapes. Obviously, there is no causal relation here; yet Bataille would claim his causal framework is as scientific as others. In terms of evolutionary history, the questioning of an unexamined Western belief in the metaphysics of presence means one should be leery of accepting knowledge by a technique that traces events backward through time; it means that the search for origins as a means of explanation is itself a problem that can and should be questioned. A rigorous questioning of underlying metaphysical foundations shakes the most basic assumptions of our linear, historical, alphabetic, and evolutionary thought. If it is possible even to question, as Derrida does, something as basic as our belief in a general archaeology—the belief that we can find the truth about things by uncovering their origins—it should be relatively easy to look skeptically upon what we think we know about sex.

Sexual stereotypes apply not only to men and women but also to our ways of speaking. The Romance languages contain nouns whose genders are fixed. In French and Spanish, for example, door—*la porte, la puerta*—is feminine. Is a door really feminine? Even in languages not so fond of choosing sides, the lack of gender may itself confer a negativity that is sexual, in light of its absence. In German, children are referred to by the neuter pronoun—here the absence of gender itself has sexual connotations. People tend to think of cats as feminine, dogs as masculine. Far more often than boys, girls have names such as Lily and Rose—the names of plants, which are considered more feminine, more "pretty" than animals. Hegel, continuing a long tradition of Western philosophy in which the male is regarded as active and ensouled, while the female is considered passive and soulless, compared women to plants and men to animals. According to many historians of culture, agriculture and the beginnings of civilization are rooted in metaphors of a fertile Mother Earth and the masculine act of planting seeds in its furrows. Mircea Eliade writes that "the divine couple, Heaven and Earth, presented by Hesiod, are one of the *Leitmotiven* of universal mythology." One southern African tribe has a marriage chant that may be translated, "The Earth is our mother, the Sky our father. The Sky fertilizes the Earth with rain, the Earth produces grains and grass." That variations on the Mother Earth theme are found among the peoples of Europe, the Americas, and Asia—from the marriage chants of Africans to the hymns of Pacific Island tribes—is suggestive, to say the least. In some languages the words for spade and phallus are identical. The Hindus thought of the agricultural furrow as a vulva, the seeds as semen. The *Satapatha-Brahmana* states, "This woman is come as living soil. Sow seed in her, men." We read that "woman is the field, and man the dispenser of seed." It would seem that the identification of earth with mother, with inseminable female, lies at the very root of Indo-European civilization, before the birth of Western philosophy and science among the ancient Greeks, contemporaneous perhaps with the origins of writing—of "history" itself—in the fertile crescent between the rivers Tigris and Euphrates in Mesopotamia, and throughout the ancient Mediterranean world. The division and union of the Sky Father and Earth Mother to make room for humanity is a mythical complex, or mytheme, so widespread that it could not simply be the result of cultural dispersion. It would be a serviceable contradiction to call it an archetype of human consciousness, separately developed in far distant communities, especially those dependent upon agriculture for their livelihood. In English, sperm and semen both come from the Latin word for "seed," referring to the seeds sown in the ground. But the connotations of this "archetypal" complex spill over far beyond a local concern with word origins. The Sky God figures of Judaism, Christianity, and other traditions cannot be disentangled from this mytheme. According to some historians, a maternal goddess religion, replete with moon worship, fertility goddesses, and menstrual calendars, preceded the male (chauvinist) religion associated with agriculture. But at the dawn of history the paternal religion prevailed. Exemplifying the alleged role reversal is the Apollo of Aeschylus' *Eumenides,* who states, "The mother is not the parent of that which is called the child, but only nurtures the seed that grows. The parent is he who plants this seed." Feminists decry these metaphysical roots of female passivity and masculine virility, which later "flowered" in Aristotle's claims that females were literally imperfect beings, mutilated males whose *catamenia,* or menstrual fluids, were impure, lacking (unlike semen) "the principle of soul." Yet, talk of "raping" the earth, while by definition inadmissible to environmentalists and feminists, partakes of the same metaphor of passive Earth Mother lamented as inappropriate and demeaning. And even if we could suddenly and radically reverse, or "cross-dress," the mytheme, make Earth a father and the Sky a mother, would we thereby escape the confines of the mytheme? In fact, since the Egyptians believed in a male earth and a female heaven, the present configuration of genders seems already to have been inverted at least once.

As if to parody the whole entrenched Mother Earth complex, Derrida in his writing makes constant use of images of sexuality in relation to writing. The style of a writer, for example, is akin to his stylus—an instrument of writing or marking, such as a hard-pointed pen-shaped instrument for marking on stencils used in a reproducing machine. Derrida uses the terms "hymen" and "dissemination" to describe the physically active, psychoanalytically indebted processes of marking paper, meaningfully or not, with the ink that flows too quickly from the pen and never enters, but closes in upon the tactile and skinlike "text." Indeed, writing, which first appeared in sedentary civilizations organized around agriculture that afforded people the leisure time to accumulate and trade goods, is deeply— or, what in this case amounts to the same thing, superficially— implicated in this riding roughshod over and leaving semipermanent tracks in the rounded, swallowing, dirty, and fertile—feminized— body of the ear(th). The cover or sheet is not only a piece of paper but a rustling sliver of bark, a marked body. As Wallace Stevens said, the poet looks at the world as a man looks at a woman. (S)he imagines.

II

It would be wrong to think that such metaphysical foundations as the Mother Earth complex are automatically escaped by the objective nature of modern science. In the first place, the identification of femaleness with femininity, passivity, and nurturing helped prevent females from even becoming scientists; and this, no doubt, biased (and still biases) medical and scientific research that is concerned with gender distinctions. The ancient Greeks, whose influence upon us remains so vast, not only accepted homosexuality but, as in the case of the Spartans and Athenians, considered the transmission of semen between males a first-rate means of pedagogy rather than pedophilia; and the *arrete* of semen, in public at the temple of Apollo (among other places) through homosexual intercourse, conferred manly nobility rather than shame. Aristotle, the inventor of biology (carrying the torch, as it were, of Socrates and Plato), who was to have great influence upon science and religion, absorbed his culture's view of the ignoble role of women. Aristotle wrote that female

infants remained unformed boys, beings whose development had
been frustrated because the colder material of the mother had cooled
off the heat of the sperm. Boy children arose from the stock of males
whose semen was sufficiently hot to remain active despite the cold
material of the mother and therefore reached a full maturity. The
influential Christian theologian Aquinas carried on Aristotle's supposedly scientific formulation that the sperm "tends to the production of a perfect likeness in the masculine sex." Today, the
supplementary nature of the female sex within Christian doctrine,
from her creation from Adam's rib to recent papal statements that
women should find their place and happiness through the love of
their husband, is well documented.

While this is not science, but religion, one must remember that
the two often shared the same bed. With the invention of the microscope, the Dutch discoverer of microbes, Anton van Leeuwenhoek,
observed his own sperm under the microscope and wrote to the Royal
Society of London that he had confirmed Aristotle. In 1685 Leeuwenhoek declared that the anatomy of his spermatazoa—a scientific
word combining the Greek words for "seed" and "animals"—
definitively proved that animals arose from semen alone. The uterus
was a mere receptacle, a nutrient medium, a container or environment
in which self-contained "spermatazoans," seed animals representing
the active male principle, developed and flourished. One might say
that Leeuwenhoek's observations led him to conclude that the ovaries
of mammals were ornamental, or that the eggs of chicken were mere
vegetable matter nourishing the animal sperm; yet, it is clear that he
was metaphysically preconditioned to perceive what he saw. Some
textbooks not much later illustrated tiny preformed men already existing and lodged within the sperm cells; this was the so-called preformation hypothesis. No one, of course, ever saw the dubious
"homunculi". The inveterately curious Leeuwenhoek examined his
own sperm with his homemade microscope; but he was uncertain
whether he could see the tiny bodies and say, "Here lies the head and
there as well the shoulders and the hips": he wrote to the Royal Society of finding "an animal whose male seed will be so large that we
will recognize within it the figure of the creature from which it
came."

Even after preformation was disproven, theories persisted about the active male principle of hungry, moving sperm as contrasted with the merely passive, receiving, and nutrifying egg. As Scott Gilbert and the Biology and Gender Study Group at Swarthmore College write in the unpublished version of a monograph originally titled "The Importance of Feminist Critique for Modern Biology," the "microcosm/macrocosm relationship between female animals and their nutritive, passive eggs and between male animals and their mobile, vigorous sperm was not accidental." They continue, "The comparison of the sperm with a vigorous active male and the egg with a passive female (to be awarded the victor) is still seen in many textbooks." Still, the writers of this feminist critique would seem to want simply to invert, or at least redress, the infiltration of the subject/object, male-acting-upon-female mode. They quote a "new account of sperm-egg interactions" in which the egg, once a "silent partner," becomes an ever more "energetic participant in fertilization." The evidence to support this role reversal comes from observations with the scanning electron microscope in which "the egg directs the growth of microvilli—small finger-like projections of the cell surface—to clasp the sperm and gently draw it into the cell. . . . Freshly ejaculated mammalian sperm are not normally able to fertilize the eggs in many species. They have to become *capacitated*. . . . The sperm and egg are both active agents and passive substrates." In *The Rights of Women* (1792) Mary Wollstonecraft tells an anecdote of a woman who asked whether it was proper for women to "be instructed in the modern system of botany": to some at the end of the eighteenth century it was a prudish question; to others, the answer simply was no. Medical correspondents have reported that in times past, proper European women had no name for the vulva and spoke to their physician merely of "lower parts" or "down there." As the self-styled psychologist of sex Havelock Ellis wrote early in this century in *The Evolution of Modesty,* "At the present time a knowledge of physiology of plants is not usually considered inconsistent with modesty, but a knowledge of animal physiology is still so considered by many. Dr. H. R. Hopkins, of New York, wrote in 1895, regarding the teaching of physiology: 'How can we teach growing girls the functions of the various parts of the human

body, and still leave them their modesty? That is the practical
question that has puzzled me for years.' " Ellis's study, which refers
to women as the (more) "modest sex," speaks of the Muslim custom
of covering the face, of the Chinese custom of binding the feet and
keeping them from view, and of the sexual allure of blushing.
Dating, coquettish flirting, courting; these all characterize the
woman, who presents herself as partially but not fully available to a
pursuing man. The courting period, in which modesty is assumed or
feigned, appears to be important for women to gauge the desirabil-
ity of men, to weed them out, as it were. The modesty of women,
their cross-cultural veiling, may have originally set the metaphysical
precedent under which women appear (to men) as a symbol or
enigma of something hidden in being. Across many cultures, one
could make a case for an identification of women with the unknown,
with a modesty in the face of the desire to know—or tell. Under
these conditions, woman as a figure of consciousness comes to stand
against open knowledge; as in the philosophy of Nietzsche, knowledge becomes masculinized, while mystery and ambiguity become
feminized, to be feared and adored as a kind of nonknowledge that
does not think *about* subjects, but simply thinks them, directly,
immanently, intuitively. This is the metaphysical bias. By means of
a kind of soundproof and *philosophical* bridal veil, the most becoming
woman, as woman, remains precisely she who does not speak or
speaks very softly, she who averts her eyes and closes herself off, a
nonverbal and nonverbalizing symbol. When the woman does speak,
she becomes, within this mytheme, more masculine. Within this
economy, obscurity and inaccessibility are as feminine as the loops
and clasps of an ornate garment.

The writers of the Swathmore draft correctly point out that botany has been sexualized into a female discipline; female and thus
unimportant, it has often been dropped from university curricula
after the converging of zoology and botany into unified departments
of biology. They quote approvingly the Minnesota botanist Conway
MacMillan, who fiercely decried the trend at esteemed institutions
such as Johns Hopkins and Harvard of offering courses not "in
biology at all, but courses in zoology masquerading under an attractive but deceptive name. Chairs of biology occupied by men

practically ignorant of one-half of the content of the science they profess to teach are not unknown in institutions otherwise altogether reputable."

As important as the feminist critique of biology no doubt is, the very division into *two*—passive/active, zoology/botany, male/female, subject/object—must be questioned. Philosophically, there is more to be done than simply reversing—or lying side by side—the mythological pair Earth Mother and Sky Father. While it is impossible to escape totally the metaphysical presuppositions that underlie scientific thinking, knowing that these foundations exist is a necessary first step in resisting the prejudices of gender-biased thinking. For example, the best current taxonomy divides living organisms into at least five kingdoms: the bacteria (kingdom Monera), the protozoans and other miscellaneous beings (kingdom Protoctista), the fungi (kingdom Fungi), and the kingdoms Plantae and Animalia. You will notice that with this scheme, the "rule of two" has been upset. If the divisions of life correspond to the archetypal division male/female, which kingdoms are male, which female? Indeed, the announcement that there are five rather than two kingdoms effectively ends the correspondence between the kinds of life and the two human genders. Either that, or it suggests that the presence of the secondary (nongenetic) sexual characteristics, such as penis, breasts, and so forth, are no longer sufficient in determining the gender to which a fellow human belongs. Men are not only opposed to but inside women. In more ways than one, the once ironclad male/female distinction begins to show signs of fatigue.

In Euripides' play *Hippolytus,* the eponymous protagonist, a worshiper of chastity, rails against women. The occasion is the absence of his father, during which his ill-advised stepmother admits she is hopelessly in love with him. In a monologue Hippolytus asks of Zeus why he deigned to make women. If you wanted us poor mortals to produce offspring, Hippolytus says, why did you not simply allow us to *purchase* them? In this we see the quintessence of the Greek attitude displayed. The metaphysical assumptions underlying Hippolytus' lament are that women are extra; superficial; supernumerary; out of the mainstream or flow of heredity, riches, power, and information that occurs along the male line. They are marginal, soulless things, whose wombs are like so many potted plants in which the seeds of personhood, that is, male souls, grow. The active growth is an animal characteristic, bordering on mobility, defining a boundary between the passive female principle in nature and the active male one. And to deconstruct these basic biases at the roots of Western metaphysics, it is not enough to wonder, as has been the recent fashion in Western biology, why males ever evolved. This is a simple reversal, a switch of the metaphysical hierarchy that has kept men on top. Like the human travelers in the science-fiction film *Planet of the Apes* who come back to earth to see men in cages and apes ruling, the turning upside-down does not upset the zero-sum relationship of power, only the positions. (Let us be clear that we are talking here of *deconstructing* the metaphysics of sexuality, in the Derridean sense. As Derrida writes in *Positions,* his method is basically one of reversal and displacement. In terms of the metaphysics of sexuality, feminist awareness has already accomplished an effective reversal of some of our assumptions. The displacement of power relations, however, has not yet occurred.)

Let us look more closely now at what has been called the maintenance of sex in evolutionary biology. Serious books with titles such as *The Redundant Male* and popular articles such as *Why Sex?* alert us to the concern within academic biology with solving the puzzle not of sexuality's origins but of its maintenance. The reasoning for this puzzle goes approximately as follows: If asexual organisms are able to reproduce perfectly well, without the energetic bother (for Hippolytus, the pure agony) of coming up, pairing, joining together each generation, why do sexual organisms remain sexual? Don't, for example, bacteria and other asexually reproducing microorganisms reproduce much faster and more effectively than sexual organisms? Shouldn't parthenogenetic animals and plants, mothers bearing only daughters, eventually supplant sexually reproducing species with males and fathers, by making them, in Cherfas and Gribbin's effective phrase, redundant? The evidence of natural history is that sexually reproducing species are wide spread. There must then, according to traditional neo-Darwinian reasoning, exist some adaptive advantage that enables sexually reproducing species to flourish, despite the advantages of speed and ease accruing to members of nonsexually reproducing species. That advantage has often been believed to be the genetic variety that sexual mingling affords offspring; in rapidly changing environments, it has been assumed, the genetically shuffled offspring of sexual reproducers should be able to adapt more readily and outcompete their more uniform asexual fellows.

━━

Let's stop and even backtrack a bit to examine some of the underlying metaphysical biases here. We must note that asexual organisms are suggestively, if unconsciously, identified with *females*. In fact, there is no good reason for this identification; it is a carryover from a (for want of a better term) "Greek" way of looking at things. In biology textbooks, in biological literature, the cell produced asexually from another cell (for example, an amoeba immediately after mitosis) is called a "'daughter" cell. Actually, the offspring is neither a daughter nor a son, since its parent was and remains asexual; it is sexless. Males cannot be redundant unless females are primordial, identified with the presumed asexual precursors to sexually differentiated plants and animals. But the prejudice was, and is, that females are less differentiated. For Aristotle, females were organisms whose entelechy had been stifled, whose telos had never been achieved. A similar teleological viewpoint tacitly assumes the resolution of evolving nature into, if not its final goal, then at least its most accomplished, reflective, and representative form, humanity. But the ugly facts murdering such beautiful teleological theories remain. *Homo sapiens* shares the earth with an estimated 30 million other living species, each as highly evolved as we. And the asexual organism is not more female than male; Adam was not, in a simple reversal, extracted from *Eve's* rib. If females are not primordial, males are not redundant.

Nonetheless, evolutionary biologists have long wondered why so many organisms are sexual when, theoretically, asexual organisms might reproduce more effectively. The traditional textbook explanation, buttressed by the mathematical analyses of population biologists such as Ronald Fisher and, later, George Williams, was that sexually reproducing species, by combining genes, enjoyed a poten-

tial advantage in changing environments over asexual species. Assuming that asexual organisms show more variation, and noticing
that some plant and animal populations can reproduce either with or
without sex, Williams, at the State University of New York at
Stony Brook, has compared the persistence of sexual reproduction to
a lottery in which the winning numbers are continually changed.
Asexual organisms, like gamblers who buy tickets with numbers
that have won in the past, but that will not necessarily win again,
lose although they may have bought a huge number of tickets.
Cherfas and Gribbin, noting the advantage that asexual organisms
have in being able to reproduce more quickly and effectively, wonder why natural selection has not produced a world of virgin mothers
and daughters, unmolested by the time-consuming rigamarole
mandated by men and sex. After all, women play the major role
in bringing up offspring and could, hypothetically, reproduce without men.

Part of the answer to this conundrum of evolutionary biology may
seem obvious: In complex organisms such as ourselves, it is not easy
to get rid of sex. Although some animals do have both sexual and
parthenogenetic species (feminist dream, or nightmare?) this does
not mean that once established, sex is easily lost. In fact, exactly the
reverse is true: Biologically, it is very difficult to break away from
two-parent sex, and so far it has been impossible for animals and
plants to lose all processes associated with it. The meiotic stage of
prophase, a sexual process, albeit at the intracellular level, remains
even in parthenogenetic animals. Although strictly daughter-
producing species of lizards and rotifers exist, the prophase part of
meiosis, in which the chromosomes line up in diploid body cells—
cells on their way to becoming haploid egg and sperm cells—seems
to be an indispensable stage in the lives of animals and plants.
Apparently, the prophase of meiosis is never lost in animals and
plants with complex tissue differentiation. It is not, therefore, sexual reproduction that is maintained, but sexually reproducing
organisms.

Moreover, uniparental organisms are not, as is commonly believed, clones. Many show much variation. Thus, another problem
with the traditional explanation of why organisms remain sexual is

that, contrary to widespread belief, sex is not required to generate adequate variation. The situation here resembles that of the amount of sexual intercourse required to generate sufficient human offspring. A "nymphomaniac" may have fewer children than a "prude": Just as beyond a certain limit, sexual intercourse becomes redundant for the purposes of generating offspring, so too, beyond a certain limit, genetic variety is redundant for the purposes of natural selection. In fact, closely related single-parent and biparental organisms both show considerable amounts of variation; entire taxa lacking two-parent sex (Fungi imperfecti and some leguminous plants) show enormous variation. The idea that sex is maintained because sexual organisms, which are presumably more varied, can adapt faster to changing environments also has been tested. Somewhat to the surprise of biologists, it turns out that the above-mentioned fatherless lizards and rotifers are just as numerous as their sexually reproducing counterparts. (Graham Bell has collected other such surprises in his book *Masterpiece of Nature.*) The further it is explored, the more this very basic assumption that asexual organisms lack sufficient variation to adapt to rapidly changing environments fails. Even single cells growing in culture by mitosis show much variation from cell to cell.

The question of how meiotic sex *originated* differs radically from that of how meiotic sex is maintained. The origin of meiotic sex is a story that cannot be taken up here; suffice it to say that once meiotic sex arose, it thrived. What accounted for its persistence? Various theories have been advanced, most based on the unproven and probably anthropocentric assertion that sexual organisms "evolve faster." Fisher first proposed that though sex is bad for individuals, because it distracts them from the business of reproducing, it has been preserved because it is good for species: In the long run, sexual species have branched off, while asexual species, their members all adapted to a single environment, foundered and became extinct during times of rapid environmental change. G. C. Williams turned this idea on its head, paradoxically suggesting that sexual species were maintained in evolution because they were poorly adapted, their constant shuffling of genes preventing them from specializing to a particular niche that would make them extinct if it disappeared.

In the 1960s in the United States the Drosophila (housefly) genet-
icist H. J. Muller suggested that sexual organisms could evolve
faster because, with two sets of genes, they would be protected from
deleterious mutations by complementary good genes. When organisms accumulated a preponderance of negative variations, they would
die out; but potentially useful genetic information would remain in
circulation unless so many mutations accumulated that they would
kill their bearers and thus effectively be dumped from the gene pool
all at the same time. In asexual organisms, however, lethal mutations could never be masked by good genes or salvaged for their
possible helpful uses with other combinations of genes in the future.
Leigh Van Valen, professor of evolutionary biology at the University
of Chicago, came up with the Red Queen hypothesis, named after
the character in *Alice's Adventures in Wonderland* who said, "It takes
all the running you can do, to keep in the same place." The Red
Queen hypothesis states that once sex evolves, the environment,
consisting largely of other organisms, jumps to a state of more rapid
change. If asexual organisms have insufficient recourse to sources of
genetic variation, they are left behind, remaining adapted only to
the less dynamic environment of the past. Viewing sex as relatively
neutral with regard to its adaptive advantage, Richard E. Michod,
of the University of Arizona, believes it has been maintained as a
method of repairing damaged genes. "Males," he says, "are a way of
providing redundant information. When females are damaged, they
can use information from males to repair their bad genes."

Common to this diverse group of theories about the reasons for
sex is the assumption that sex contains some sort of advantage. But
does it? Indeed, the question, What maintains sex? may be a poor
one. It appears that microbes, such as Stentor, probably have lost
sex; while this organism will occasionally attempt to mate, mating
inevitably results in death—clear evidence that sexuality may be an
atavism, like the human love for salt and sugar in an age of easily
available junk food. But sexual reproduction may be far more difficult to get rid of in more complex and highly orchestrated cell
collectives, such as mammals. If the prophase stage of meiosis is
indispensable, it cannot be eliminated unless the organisms themselves are eliminated, too. In many complex organisms, the meiotic

stage of prophase is tightly integrated into an entire life cycle that includes everything from flirtation to mating rituals and physiological processes involved in the releasing of eggs and the making of sperm. Examining the assumptions underlying the question "Why is there sex?" may reveal the question itself to be flimsily erected. Paraphrasing Robert Kennedy, we might say, regarding the persistence of sex, "Some men ask why; I ask why not."

III

Here, metaphysically, sex is a certain bottomlessness that cannot be categorized. Maybe, then, it is good that we cannot be objective when it comes to sexual matters. If sexuality inscribes us in the core of our being, we *are* insofar as we are sexual beings. The body that each of us possesses would also possess us. For the Greeks the things of Aphrodite, *aphrodisia,* combined the sexual desire leading to sexual acts and the sexual acts leading to sexual pleasure with the sexual pleasure leading to sexual desire. According to Foucault, although the theme of sexual austerity apparently remains constant within the history of sexuality, pagans were sexually temperate for different reasons than were Christians. The Greeks tended to resist temptation as a kind of bodily regimen, a mastering of an aphrodisiac cycle that might otherwise snowball out of control. A general "diaetetics," or "aesthetics of being," prevailed among those who wrote; the morality of self-mastery had more to do with achieving domination of self and others, but this ascetic ideal (which applied also to food, sleep, and wine) metamorphosed, in Christian times, into a morality of purity based on bodily integrity rather than self-restraint. The Greek writers were, apparently, less concerned to lay down specific rules for appropriate and ethical behavior, than to awake in the literate elite a sense of rightness of character. They focused less on prohibition, on who could do what with whom, than on *aphrodisia* and *chresis,* the regulated use of pleasures that were not in themselves forbidden.

With pleasure and passion, in agony and ecstasy, we think, perceive, grow, live, feel, make love, and die. Each person, each subject, is involved in the world in a physical relationship of subtle sensations and communications that no popular or scholarly account could ever completely encompass or describe. The world, the body, *are* us and yet they tempt us, too; dissatisfaction can be brutal or ethereal, partial or impartial, dependent and transcendent, impish, playful, possible, curious, and experimental. The interplay of genes creates new individuals in the biological realm, the recombination of letters creates new words, the recombination of words creates new thoughts, the recombination of colors creates new paintings, and the recombination of materials creates new technologies—sometimes. Metaphysically, sex is bottomless, a transcendental game of recreation.

Is sex then part of a broader connecting phenomenon, a sort of parasexuality that brings together not only organisms of the same species but those of different species, as well as objects, molecules, friends, and ideas? An old theory holds that one of the central features of evolution is the increase over time in the rate of concentration and dispersal of chemical compounds. The flux of matter at the surface of the earth, the movement together and apart of atoms, organisms, and ideas is increasing. The attracted and repelled, the opposites and likenesses, all join together in a wild swirl. Scientifically, an alluring young woman may be only a temporary combination of nucleic acids, fats, proteins, and carbohydrates—of leather, lace, and lipids—but for a young man the ineffable mystery of her charm always remains.

When bacteria reproduce, they do not require genetic exchange, the swapping and donating, the receiving and taking of genes in the form of nucleic acids. Nonetheless, transformation, transduction, and other bacterial gene-trading processes are as old as life itself. It even has been proposed that sexuality preceded the evolution of the first form of life, bacteria. Bacterial sexuality, which is not a prerequisite for bacterial reproduction, is hypothesized to have evolved in the earth's early environment under conditions of harsh ultraviolet radiation from the sun. The solar radiation was more intense 3.5 billion years ago, when bacteria evolved, because the atmosphere had no ozone layer. (Ozone is a by-product of free oxygen, which had not yet been released into the atmosphere, since cyanobacteria—

photosynthetic organisms that release oxygen atoms from water—had not appeared.) Bacteria bombarded by ultraviolet radiation survived by grafting copied pieces of DNA onto themselves to replace DNA damaged by ultraviolet radiation. When pieces of DNA from one organism worked their way into pieces of another organism, sexuality, in its strict biological sense, evolved.

"These scientists don't know when to stop," preached the voice over the radio. The electronic pulpit belonged to a Christian broadcasting station in Florida, but it was protesting a meeting taking place hundreds of miles away, in Woods Hole, Massachusetts. The preacher was objecting to a meeting entitled "The Origins and Evolution of Sex." "When I was young," said the man, "they left it at the birds and the bees. You would think they would be satisfied to leave it at the level of the flea. But that is what happens when you don't follow the Word of God. These scientists have to push it further, to bacteria—to germs. It's kind of disgusting. If they want to know about the evolution of sex, it's all written there in Genesis, in the story of Adam and Eve. God himself has told us in his own words."

The insertion of the preceding paragraph was an attempt to add novelty to this essay. It does not really belong here, and yet it wants to be integrated. The origin of sex in bacteria may have been a very similar process, insofar as the inclusion of extra information was originally accidental. In some cases, however, the added genetic information conferred new qualities upon the bacterium that helped it survive in conditions that otherwise would have been impossible. For example, strains of Gonococcus, the bacterium responsible for the venereal disease gonorrhea, have evolved that are immune to penicillin. This immunity was conferred upon Gonococcus by penicillin-resistant bacteria via bacterial genetic exchange—that is, bacterial sex. The bacterial sex may well have occurred within the intestines of a human being: We often take antibiotic preparations, with the cumulative outcome that many of the bacteria normally inhabiting our gut develop a resistance to bacteria. The genes of one of these, intact and traded to the disease bacterium, accounts for the latter's resistance to the antibiotic penicillin. Here, in the zone of the unforbidden, the pen is—one might say, im possible—mightier than the sword, in the extended sense of lyzing and analyzing, of cuttings preliminary to gluing and regrafting. A prosthesis billions of years old.

The discovery of transposable elements—"parasitic DNA," which is neutral and irrelevant but manages to copy itself within the genetic endowments of organisms—has fostered much speculation among some biologists. Indeed, according to the molecular biologists Michael Rose and Ford Doolittle, sex "might reasonably be seen as a form of disease that animals and plants have learned to live with." They postulate that sex is a kind of inherited misfortune, that "males can be seen as parasitic DNA made manifest at the organismal level." Present-day bacteria, exposed to ultraviolet radiation, will let loose plasmids; that is, they will explode into genetic fragments. These plasmids may be absorbed by other bacteria, the genes of the plasmids providing, in some cases, a benefit. Still the orgiastic breakup should be seen more as the legacy of ultraviolet radiation on the early earth than as a concerted response, a way of adapting in biological crisis. In the same manner that plasmids sometimes fortuitously endow neighboring bacteria with useful genes, so males sometimes prove beneficial to females. But in the main, they are, according to Rose and Doolittle, a pain, a necessary bother. The whole scenario has little to do with the view among ancient physicians that sexual passion is akin to disease, as attested to by the "madness" of lovers and their "spastic" nature in bed together. Rose and Doolittle's view also has little to do with performance artist Laurie Anderson's song in which she slowly intones, borrowing a line from William S. Burroughs, "Language—language is a virus."

The wanton movement of elements, the reintegration of extra and copied parts to make new wholes, is a plagiarism as old as life itself. It confers upon the bacterial world the nature of a planetary superorganism with cells woven over the entire globe, while on the scale of modern communications it provides for international cultural transfer, the modular storing and creation of new artistic genres, technological inventions, and religious and philosophical worldviews. As biological structure (flesh, bones, bark, and so on) is coded for by the peninfinite arrangement of only a very limited alphabet of four nucleotides, so meaning itself arises from addition and attraction, the mixing of discrete units, such as letters, graphemes, phonemes, mythemes, pictures, and words. The reproducibility of any informational structure allows it to be severed from its context and, for better or worse, reintegrated into another. Such recombination may be a forced and short-lived encounter, destined to disintegrate, or it may be the basis for an enduring marriage—a marriage in which the partners ultimately meld into each other and reproduce as an inseparable symbiotic unit. Sex at the most abstract level concerns not plunging genitalia but a recombinatory and joyful spirit, an energetic play. This sort of metaphysical sex takes its rightful place not so much between throbbing bodies (though, of course, it can) as in one's own imagination. It is a sex in but not of the world. It comes before flesh and time. It can do anything words can do, because it partakes of their recombinatorial nature. Let us combine letters and call this sex an asexual or nonsexual sex. The power of one little *a* that we add in order to subtract, but which does so much more. And at the end of this essay, as seems fitting, you will see that I have put a colon:

READING LIST

Bataille, Georges. "The Solar Anus" in *Visions of Excess: Selected Writings, 1927–1939.* Ed. Allan Stekl. Trans. Alan Stekl, Carl R. Lovitt, and Donald M. Leslie, Jr. Minneapolis: University of Minnesota Press, 1985.

Beldecos, A. S. Bailly, S. Gilbert, K. Hicks, L. Kenschaft, N. Niemczyk, R. Rosenburg, S. Schaertel, and A. Wedel. *The Importance of Feminist Critique for Modern Biology.* An unpublished, unabridged version of the manuscript. The Biology and Gender Study Group, Swarthmore College, Swarthmore, PA 19081.

Bell, Graham. *The Masterpiece of Nature: The Evolution and Genetics of Sexuality.* Berkeley, Calif.: University of California Press, 1982.

Cherfas, Jeremy and John Gribbin. *The Redundant Male: Is Sex Irrelevant in the Modern World?* New York: Pantheon, 1984.

Derrida, Jacques. *Positions.* Trans. Alan Bass. London: Althone Press, 1981.

Ellis, Havelock. *Studies in the Psychology of Sex,* vol. 1, Part One. New York: Random House, 1942.

Halperin, D. M. "Sexual Ethics and Technologies of the Self in Classical Greece," review of Michel Foucault, *L'usage des plaisirs, Histoire de la sexualité, 2. American Journal of Philology* 107 (1986), pp. 274–286.

Halvorsen, ed. *Origins and Evolution of Sex.* Woods Hole Marine Biological Laboratory.

Rose, Michael, and Ford Doolittle. "Parasitic DNA—the Origin of Species and Sex." *New Scientist,* 1985, vol. 98, no. 1362, pp. 787–789.

Sonea, Sorin and Maurice Panisett. *The New Bacteriology.* Boston: Jones and Bartlett, 1983.

On the Method
of Theoretical Physics
(Redux)

GERALD FEINBERG

. . . the process of solving equations has to a large extent been discarded, at least for equations that may be regarded as fundamental. In fact, because physicists no longer emphasize solving equations, it is unclear what the fundamental equations of present-day physics really are. The process of solving equations has been replaced by something much harder to describe, perhaps because the procedures being followed are not yet completely well defined.

If you want to find out anything from the theoretical physicists about the methods they use, I advise you to stick closely to one principle: Don't listen to their words, fix your attention on their deeds.

—*Albert Einstein, in a lecture at Oxford, June 1933*

Although Einstein entitled his lecture at Oxford "On the Method of Theoretical Physics," as I have entitled this essay, he actually said almost nothing about method at all. Instead he examined the development of the basic ideas of theoretical physics. Nevertheless, following Einstein's injunctions is one of the best guides to procedure in science, and that is what I propose to do here. I will discuss the development of the method of theoretical physics over the last fifty years. My analysis leads me to believe that a profound change has occurred in the method of theoretical physics, a change hardly recognized either by theoretical physicists themselves or by the philosophers and historians who comment on their work.

What is this change? At one time, the main activity of theoretical physicists was formulating and solving equations. From Euler's mathematical expression of Newton's second law through the electromagnetic equations of Maxwell, Einstein's field equations of general relativity, the Schroedinger equation of nonrelativistic quantum mechanics, and Dirac's relativistic equation for the electron, the core achievements of the greatest theoretical physicists could be summarized as the formulation and solution of equations. These equations typically describe how some quantity, such as an electric field, varies in time. But physicists have proposed no new fundamental equations for a long time. Especially in the most basic branch of physics (which until recently was quantum field theory and has now perhaps become quantum string theory), the process of solving equations has to a large extent been discarded, at least for equations that may be regarded as fundamental. In fact, because physicists no longer emphasize solving equations, it is unclear what the fundamental equations of present-day physics really are. The process of solving equations has been replaced by something much harder to describe, perhaps because the procedures being followed are not yet completely well defined.

In certain areas of theoretical physics, of course, solving equations does remain the characteristic activity. These include all the areas that I mentioned earlier connected with the work of the historical leaders of the discipline. But for the most part, these fields are not at the frontiers of physics. They are now mostly applications of well-known ideas to novel situations.

I am not saying that theoretical physicists who work on fundamental questions, such as the behavior of subatomic particles, do not perform calculations. It is doing calculations that makes them theoretical physicists. Rather, my point is that the calculations physicists now do cannot usefully be described as solving equations. Instead, contemporary physicists focus on certain quantities that are related (closely or distantly) to observations, and use one of several algorithmic procedures to determine these qualities. The theories that are the basis of their calculations are in some sense defined by the algorithmic procedures they use, rather than by some underlying set of equations or other general principles. The heroes of contemporary theoretical physics are not those who have discovered new equations, but those, such as Kenneth Wilson, who have found new quantities to calculate or new ways to calculate. This situation can be viewed as an ironic realization of the remark Einstein made over fifty years ago in his speech.

Several developments over the past fifty years within theoretical physics have resulted in the present situation. These include new techniques for calculating the answers to physical problems, new conceptions of what is interesting, and new perceived limits on what theoretical physicists can hope to calculate. In order to see what has happened, let us compare the situation in fundamental theoretical physics fifty years ago with that today.

The earlier situation can be inferred by looking at some of the standard books published at that time, such as Dirac's *Quantum Mechanics* or Wentzel's *Quantum Field Theory*. Fifty years ago theoretical physicists imagined that they either had or could formulate a fairly simple set of equations that would describe all natural phenomena. These were a set of coupled partial differential equations describing how certain functions, the quantized fields associated with a small number of subatomic particles, varied in space and time. Furthermore, theoretical physicists saw no obstacles in principle to solving these equations, and hence specifying their consequences for any desired situation, although there certainly were practical obstacles to doing so. For the most part, the solutions were thought to be related in a fairly direct way to things that could be observed about the particles, such as their mass, numbers, and motion. The procedure in standard texts was to present arguments leading up to the equations, and then to discuss how to solve them. Usually this was done by some type of infinite series expansion, but the method was basically analytic, since scientific computing machinery was barely a dream in Turing's mind.

Several factors have changed these attitudes, in a way that I think is irreversible. One is that physicists have come to realize that in most situations the equations are simply too hard to solve effectively. Although the equations of quantum field theory resemble those of classical field theories (such as that of Maxwell), these resemblances in form mask serious differences. The quantities entering into the quantum field equations are actually not simple functions, but what mathematicians call operators, generalizations of finite matrices. Consequently, even in the absence of some external source of the field, one cannot take the solution to vanish in some region of space and time, as one could for classical fields. The absence of a field everywhere is *not* a solution to the equations of quantum field theory. Even a very simple situation, when no particles are present, corresponds to a highly complex expression for the field—indeed, one for which an explicit and exact solution cannot readily be obtained, although we understand many of the properties that such solutions must have. Furthermore, the equations of interesting quantum field theories involve several fields all interacting in a nonlinear way. Thus, it is usually impossible to find solutions in the form of integrals over prescribed distributions, as is often done, for example, with the equations of classical electromagnetism. Because of these problems, physicists have spent little effort in finding explicit solutions to the equations of quantum field theories, except for artificially constructed examples of such theories, to which they often refer disparagingly as "toy models." What theoretical physicists have done to some extent in recent years is to look instead for solutions to the corresponding equations of classical field theory. If these solutions can be found, they can be used to make inferences about the solutions to the equations of the quantized theory, by means that I will describe below. But, of course, that is not the same thing as solving the quantum equations themselves.

Several developments in the period directly following World War II contributed further to the change in theoretical physicists' attitude toward equations. First was the invention of techniques for doing certain calculations that enabled physicists to obtain definite answers for measurable quantities much more rapidly than they could with previous methods. These techniques, often associated with the use of pictures known as Feynman graphs to represent the physical process being studied, were introduced by Richard Feynman in the late 1940s and, following a short period of incredulity over their applicability, were rapidly adopted by the vast majority of younger theoretical physicists. The essence of Feynman's method is a way of associating one or more pictures with any physical process, and a series of calculational rules for determining from these pictures a quantity called the scattering amplitude. When two subatomic particles collide, different events can occur: The original particles can emerge with new velocities; extra particles can be produced; and so on. Each event has some probability of taking place. The scattering amplitude is a mathematical quantity whose square gives these probabilities. In principle, it is possible to write equations for which these scattering amplitudes are solutions. However, in practice, little or no attention is paid to those underlying equations, and the calculational techniques for finding the scattering amplitudes have come to define the physical theories.

Although some opposed the takeover of theoretical physics by

Feynman diagrams, especially Julian Schwinger and his students, the victory of Feynman's method was ensured by the increasingly prevalent attitude that in subatomic physics, the "only" observables are the results of scattering experiments. This attitude derives from the undoubted fact that most experiments on subatomic particles make no attempt to follow the development of the motion over the short intervals that elapse between the beginning and the end of the interaction. Instead, such experiments measure some quantities related to an incoming beam of particles, and some quantities about the outgoing beam, leaving what happens in between to be inferred as nearly as possible from that information.

Feynman himself first came upon the rules for calculating with diagrams by an entirely different approach, known as the method of path integration, which in the past twenty years has taken on renewed vigor and has become a new rival to the solution of equations. It is essentially a way of expressing the time development of some system described by quantum mechanics in terms of a type of integral, which involves quantities that would describe the same system in classical (that is, nonquantum) physics. In simple cases, these quantities are possible paths that the system could follow according to classical physics. So, for example, if the system is a single particle acted on by a force, then the mathematical object that summarizes the quantum properties of the particle, its Schroedinger wave function, can be related to an integral over paths in which the motion of the particle is described by Newtonian mechanics. If the system is the quantized electromagnetic field, then its quantum properties are related by a path integral to the solutions of Maxwell's equations of classical electromagnetism.

As conceived by Feynman, path integrals were an alternative way to solve the equations of quantum mechanics. His early discussions of them were, in fact, based on work by Dirac that was well within the framework of solving those equations. But, as often happens in physics, what started out as a variation on old ideas eventually came to be seen as a new idea capable of supporting greater structures of thought than was previously possible.

That development took another twenty years after Feynman's original work. The stimulus for it was the realization, by Feynman himself and others, that in the case of certain interesting quantum field theories, including that describing gravity, the rules for calculating scattering amplitudes, which could not readily be inferred by conventional methods, could be more easily obtained by the method of path integrals. This recognition gave new vigor to the use of path integrals, which became the preferred way to translate a classical field theory into the language of quantum theory. But with that change also came a significant change in emphasis. In the path integral formulation, the physicist is not attempting to solve a basic equation. Rather, the path integral is something like a generic solution to an unspecified equation, and one is trying to calculate, or at least to approximate, that solution. What is important for the path integral method is the specific solution, not the unknown equation.

The path integral for a given theory is itself specified by a quantity, known as the action, that characterizes a specific system. This quantity also played a role in previous versions of quantum field theory where it was regarded as an auxiliary quantity whose main function was to help determine the correct equations for the fields. Now, however, calculating the action in some accurate approximation has become the main focus of the path integral formalism.

Some efforts to calculate the path integral have led to another important change of emphasis. The equations of physics generally involve three space coordinates and one time coordinate, all of which are taken as real numbers. But in calculating path integrals, it is often convenient to treat the time variable as an imaginary number, restoring it to its real, "physical" value only at the end of the calculation. When such imaginary time is used, it is frequently possible to relate the calculation of a path integral to calculations that are done in statistical thermodynamics, where imaginary time is replaced by temperature. This replacement has been especially useful in some recent approaches to fundamental physical problems that involve many numerical computations. However, this procedure has sometimes resulted in treating the results obtained using the imaginary time coordinate as if they had a significance beyond that of a calculational device. Of course, that kind of transmutation occurs over and over in theoretical physics. Fields themselves began as a means of calculating forces on particles and eventually came to be seen as an independent reality. It is too early to know whether the same will be true for imaginary time.

The path integral method does involve solution of some equations: the equations of classical field theory with the same action as quantum field theory. A favorite method for evaluating the path integral has become to find some specific solution of the classical equations, compute its contribution to the path integral, and hope that this contribution dominates the actual integral. While this technique has led to some interesting ideas about possible new phenomena, it cannot be considered a systematically reliable means of calculation, since it is rarely clear that the assumed terms are truly dominant. Furthermore, in most cases, the classical solutions themselves are not thought of as representing actual physical situations, especially since they tend to be valid only when the imaginary time coordinate is used. Nevertheless, theoretical physicists tend increasingly to write and speak as if these classical, imaginary-time solutions were more than a calculational device, and actually appear in the phenomena.

It is somewhat ironic that Richard Feynman was the person most responsible for diverting theoretical physicists from the method of inventing and solving equations. Abdus Salam tells a story about the first meeting between Feynman and Dirac, which took place at the Solvay Conference in 1961, long after Feynman had done work on diagrams and path integrals. In the course of the meeting, the following conversation occurred:

FEYNMAN: It must have felt good to have invented that equation.

DIRAC: But that was a long time ago. What are you yourself working on?

FEYNMAN: Meson theories.

DIRAC: Are you trying to invent a similar equation?

FEYNMAN: That would be very difficult.

DIRAC: But one must try.

Evidently, Feynman himself was not satisfied with the form of his contribution to theoretical physics, and longed to achieve something closer to the ideal of previous generations of physicists.

▭

Solving equations has been largely abandoned in quantum field theory for another distinct reason as well. Physicists have known since the early 1930s that severe mathematical problems result from trying to solve the original equations of quantum field theory. These problems have several causes. In quantum field theory, we usually assume that subatomic particles, such as electrons, are mathematical points. This assumption leads to the possibility of the particles' exchanging arbitrarily large amounts of energy. Because of this, and because whenever particles interact, new particles can be created, the description of even the simplest subatomic particle is extremely complex. Another source of the problem is the assumption that physical equations must remain the same in all regions of space and time, regardless of the quantities involved. As a result, the wave function for fields that interact cannot be expressed mathematically in terms of the wave function for noninteracting fields. Moreover, we must conclude that the properties of interacting particles differ from those of noninteracting particles by mathematically infinite quantities.

Most of these problems in quantum field theory are given the generic name "divergence difficulties." For a while, in the 1930s, physicists tried to change the equations to avoid these problems. They found various ways to do this, but the new equations generally had problems of their own, which were as bad as those of the original equations. Gradually, physicists abandoned those efforts in favor of a different approach. This new approach, "renormalization," has been the main weapon that physicists have used since the late 1940s to make sense of quantum field theories.

The essence of the renormalization method lies in the recognition that certain quantities that occur in the equations of quantum field theory, such as the masses and electric charges of particles, are not directly accessible to experiment. The measurable values of these quantities could, in principle, be related to the numbers appearing in the equations, but they are not identical to them. For example, in the equation for the quantized field describing electrons, a number appears that is called the electron bare mass—that is, the mass that an electron would have if it did not interact with other particles, such as photons. The measurable mass of an electron differs from the bare mass because the particle that we call an electron produces electromagnetic fields through its interaction with photons, and these fields contribute to its mass. Unfortunately, the difference between the bare mass and the measured mass, when calculated by the rules of quantum field theory, turns out to be an infinite number. Since the measured mass is finite, the bare mass must be infinite. This is one of the divergence difficulties referred to earlier.

Through renormalization, certain observable quantities (scattering probabilities), can be expressed in terms of the measurable mass and other measurable quantities, known as coupling constants, that prescribe the strength of the interactions, rather than in terms of the infinite bare quantities. It turns out that for some, but by no means all, quantum field theories, the scattering probabilities can be expressed as a power series in the coupling constants, and each term in the series is a finite number. We do not know whether the series converges, although certain arguments suggest that it does not. The use of the measurable masses and coupling constants does not involve any loss of predictive power for the scattering probabilities, because the renormalization procedure expresses them in terms of one set of numbers instead of an equal set of different numbers. But in each case, the numbers themselves have to be obtained by measurement. The bare masses and coupling constants can be inferred from experiment only through certain limiting procedures, and the fact that these limits are infinite does not by itself imply anything unphysical about the theory.

The renormalization method has been very successful in several respects. As a means of calculation, it has proven capable of obtaining extremely precise results for measurable quantities such as atomic energy levels. Perhaps more unexpectedly, renormalization has turned out to be a valuable means for choosing among the many possible quantum field theories. The principle that only theories amenable to the renormalization method are acceptable has singled out a class of quantum field theories based on a principle called local gauge invariance, and calculations in those theories, using the methods of Feynman diagrams, have proven extremely successful in describing a wide variety of phenomena. The current so-called "standard model" of subatomic particles is based essentially on a combination of such gauge theories, as are many attempts to go beyond that model.

In spite of these successes, physicists have criticized the renormalization method on many grounds. For one thing, the method does not begin with meaningful equations that can be solved by standard mathematical procedures. The equations of quantum field theory still involve infinite quantities, which the renormalization method allows us to manipulate successfully to obtain finite answers for some observables. Their meaning is defined by the renormalization procedure. Often, the renormalization method is carried out not on the original equations containing infinite quantities but on modified equations that would apply to a universe that was finite in space and time, in which points could not be arbitrarily close together. The quantities calculated in this way depend on the size of the spacetime and on how close together points are allowed to be. Only after renormalization is done is the answer converted back to the form it has in our universe, which is presumed to be infinite and continuous. In view of these convoluted procedures, it would seem that the original equations are no more than mnemonic devices to remind physicists what calculations should be done. Theoretical physicists brought up in the old ways of doing things, such as Dirac, have wondered aloud and in print whether procedures with such weak foundations can be trusted, whatever their apparent success.

Second, renormalization does not give finite answers for all quantities that are defined within quantum field theory. The method is designed to give finite answers for the scattering cross-sections that are thought to be the ultimate observables, but in many renormalizable theories, including those in the standard model of subatomic particles, other quantities, which in previous approaches to physics were considered as basic as the scattering cross-sections, do not come out to be finite even after renormalization. These quantities include the function that describes how a system evolves from one finite time to another. This situation might be satisfactory if the arguments implying that scattering cross sections are the only observables were entirely convincing, but such arguments are usually rather vague and heuristic—in contrast, for example, to some of the analyses of measurement that have been given in the context of ordinary quantum mechanics. It is not completely clear that everything that might possibly be observed is really calculated to be finite through the use of renormalization.

Finally, the renormalization method does not work for all quantum field theories, in the sense that it does not lead to finite results even for quantities that are known to be observable. It is applicable only to some theories, such as that describing the interaction between electrons and light. As I have already remarked, this fact has been used as a tool for choosing among possible quantum field theories—for example, the theory that unified electromagnetism and weak interactions is renormalizable, while some of those that it replaced are not. Unfortunately the quantum theory of gravity appears to be nonrenormalizable, which implies that some approach going beyond renormalization is required to include the effects of gravity in particle physics.

Partly for this reason, many physicists have been attracted to a new type of theory, in which particles are replaced by objects of very small but finite size, known as strings. Because the strings are extremely small, there are few differences between the two types of theories for presently known phenomena. But the string theories seem to have the possibility of giving finite results for all observable quantities, including those involving gravity, without the specific need for a procedure such as renormalization. However, at least at present, string theories also have some of the negative features of field theories; they are characterized by procedures for calculation rather than by well-defined fundamental equations, and it is unclear whether they really give finite results for all quantities of physical interest.

What is the ultimate significance of the change I have been describing in the methods of theoretical physicists? While it is easy for someone raised in one scientific tradition to criticize something new, I do not think that such a change is necessarily to be condemned. Nothing about formulating and solving equations makes that method of obtaining information about nature intrinsically better than others. Its use in physics is no more than a few hundred years old, and may plausibly be superseded someday as were earlier methods. But at least two aspects of the current practices of theoretical physicists do need to be criticized and ultimately corrected. One is the failure to recognize the indicated change in what we are doing. It is unhealthy for our branch of science that we regularly invoke principles that we no longer really follow. We are a bit like a man who happily sins all week, and then expresses allegiance to the Ten Commandments in church on Sunday. This problem is especially acute in regard to communication of the procedures we use to others not already aware of them, as in the training of new physicists.

Also lacking in the present method of theoretical physics is a careful formulation of the ultimate principles underlying physics as we now conceive it. So long as we had fundamental equations, these equations could be taken as defining such principles. In the nineteenth century, physicists were confused about whether a mechanical model was needed to explain the electromagnetic theory of Maxwell. The situation then was decisively clarified by the remark of Hertz: "Maxwell's theory is Maxwell's equations." But if we no longer feel the need for such a defined set of equations, it is unclear what our physics is based on. It will not do to use a set of calculational procedures as a basis, although such a basis may evolve out of these procedures, somewhat as quantum mechanics evolved out of the earlier procedures of Bohr, Sommerfeld, and Ehrenfest.

There are several reasons for rejecting calculational methods alone as a suitable basis for physics. One is the lack of a guarantee that any specific set of procedures satisfies even minimal criteria of consistency with general physical principles. When Feynman first stated his rules for calculating scattering amplitudes, those rules contained an inconsistency with the conservation of probability, which was found only by comparison with alternative methods based on the use of the equations of quantum field theory. Even today, for some of the gauge field theories, it is uncertain what rules should be used to obtain answers for scattering amplitudes that are consistent with such requirements as relativistic invariance.

A more fundamental reason for rejecting calculational procedures as an ultimate basis for physics is that such procedures are effective only to the extent that we know in advance what we want to calculate. In that situation, the desired result drives the calculations. On the other hand, when equations or some other underlying principles are available, these principles determine what should be calculated. When that is done, the equations have often been found to imply the existence of types of phenomena that were previously unsuspected. These novelties would probably not have been recognized if theoretical physics had been limited to some specific calculational method for phenomena that are already known.

One reason we have recently been content with means to calculate is that for most of the past forty years or so, fundamental theoretical physics has been trying to catch up with the wealth of phenomena that experimental physicists have observed, mostly through experiments in which subatomic particles collide with one another. This has led to the emphasis on finding means for calculating what is observed in these collisions. But in the last few years, there have been few new discoveries in particle physics. It is perhaps this change that has rekindled physicists' interest in finding a new basis for describing nature, a quest of which string theory may be just the first step.

It is interesting to speculate about a new basis for theoretical physics. I doubt that it will be some new equation, which physicists will try to solve as in former days. I think it more plausible that investigation of structures rather than spacetime evolution will become the basis of theoretical physics. That is, theoretical physics will seek to identify the universe with some specific mathematical structures whose detailed properties are worked out by a wider variety of methods than just the solution of equations. An unsuccessful past example of what I have in mind was the effort by Kepler to understand the distances of the planets from the sun by regarding their orbits as inscribed inside the five Platonic solids. This approach involved calculations, but not the solution of equations. Of course, because of the progress of mathematics, we now know of many more structures, such as exotic topological spaces, that might be used to describe the physical universe. If my speculation about the direction of theoretical physics is correct, many more possibilities will emerge for physicists to explore. My impression is that mathematicians study these structures mostly through the use of methods other than the solution of equations, and I expect that physicists will follow in that direction as well.

How to Tell
What Is Science from
What Isn't

RICHARD MORRIS

*If a theory is crazy, or unorthodox, or seemingly bizarre,
that does not make it pseudoscientific. Crackpot and pseudoscientific theories are bizarre in a particular way. They
tend to ignore long-established scientific ideas. They operate in a world of their own, not in the world of scientific
discourse.*

I think it is safe to say that most scientists feel it isn't really difficult to tell what is science and what isn't. Physics is science. So are chemistry and biology and geology. Astrology isn't science. Neither is Dianetics or creation science. Stephen Hawking's theories about black holes, these men and women conclude, are scientific, as is research into the genetic code, or the study of superconductivity. On the other hand, it is generally assumed that the claims of UFO investigators, studies of psychic phenomena, and books about crystal healing and perception in plants are not.

The distinction seems to be fairly straightforward, at least until someone asks, "But how *do* you distinguish the scientific from the nonscientific?" Then it immediately becomes apparent that it is not so easy to come up with a good answer.

When asked about nonscientific or pseudoscientific ideas, scientists generally reply, "Oh, there's no evidence for anything like that," or something similar. They point out, for example, that it is difficult to believe in the existence of UFOs when no artifact of extraterrestrial origin has ever been found, when the only "evidence" we have for the claim that the earth is being visited by alien craft consists of a few blurred photographs and the tales related by people who say that they have been abducted and sexually molested by strange creatures.*

The only problem with this argument is that it is not very convincing. After all, scientists also speculate about ideas that are, as yet, unconfirmed by empirical evidence. If they didn't, scientific progress would come to a halt, or at least slow considerably. We could make a good case for the proposition that there is more evi-

* The sexual molestation theory is elaborated in detail in Budd Hopkins's *Intruders* (New York: Random House, 1987).

dence for astrology than for some of the ideas currently being pursued in the field of theoretical physics.

For example, an idea that is currently very fashionable among theoretical physicists is the notion that certain hypothetical objects known as superstrings may prove to be the ultimate constituents of matter. According to the theories currently being developed, all subatomic particles are made of superstrings. There literally isn't anything else. It is true that if any of these theories turns out to be correct, a great deal will be explained, and a number of outstanding problems in physics will be cleared up. A superstring theory could very well turn out to be the long-sought "theory of everything" upon which all other physics could be based. If we knew what the universe is made of, we wouldn't be far from being able to explain how it works.

This line of attack has only one problem. No one has ever seen a superstring, nor does anyone hope to observe one in the foreseeable future—indeed, superstrings may never be observed experimentally. The problem is the size of the superstrings, which, if they do exist, must be extremely tiny: The relationship in size between a superstring and an atomic nucleus would be about the same as that between the nucleus and the Earth.

Incidentally, we can "see" atomic nuclei; although we can't photograph them, obviously, we can detect them in a number of ways. One of the most straightforward consists of bombarding nuclei with subatomic particles and watching these particles bounce off, which is how the British physicist Ernest Rutherford discovered the atomic nucleus in 1911. Naturally, he had no way of observing his subatomic projectiles directly; to get around that difficulty, he simply set up fluorescent screens, and the particles produced points of light when they struck them. A television picture is created in a similar way; light is produced when electrons strike a fluorescent coating on the inside of the picture tube.

It is also possible to see particles that are smaller and lighter than atomic nuclei, though there are limits, because the smaller an object is, the greater the amount of energy required to see it. To create more energy, it is necessary to build larger and more expensive particle accelerators; this is the rationale behind the "supercollider" that is to be built in Texas.

Unfortunately, no matter how many millions of dollars are spent on the supercollider, no one will be able to use it to detect the existence of superstrings, which would be much too small. We couldn't hope even to see them with a particle accelerator the size of our solar system.

Superstrings are thought of as vibrating loops in ten-dimensional spacetime, which may be conceived as one dimension of time combined with nine of space (physicists use the term spacetime since Einstein's discovery that it is mathematically simpler to consider space and time together). The extra six dimensions required by superstring theory are presumably curled up into themselves so tightly that it is as impossible to detect them as it is to see the superstrings. Some relatively simple analogies make this "curling up" process easy to visualize. Think of a circle as a straight line that has somehow been curled up upon itself; then imagine the circle's being curled up more and more tightly until it is indistinguishable from a mathematical point. Alternatively, a sheet of paper can be curled up into a cylinder, which can then be wound up more and more tightly until it begins to resemble a moderately thin rod. In each of these cases, a dimension of space seems gradually to disappear. In superstring theory, the same thing happens to the six extra dimensions (if, indeed, there are only six; some versions of the theory require even more). No one knows whether these six extra dimensions of space exist; no one knows whether superstring theory even requires that they be there. Theoretical physicists do not yet agree whether the extra dimensions are to be thought of as real or only as convenient mathematical fictions.

Such ideas sound even more bizarre than the tales told by UFO abductees, yet we don't hesitate to call such speculation "scientific." It appears that we cannot distinguish true science from fringe science, pseudoscience, and assorted crackpot ideas, simply by saying that there is no evidence for the latter. At this point, it may be useful to examine the case of astrology. Astrologers claim that theirs is an empirical science. They say that when they interpret birth charts, they are making use of correlations between planetary aspects and events in people's lives that were observed in ancient times. For example, a conjunction between Jupiter and Saturn is supposed to have a certain meaning because ancient astrologers noted that this configuration influenced people's lives or personalities in predictable ways.

We generally reply to astrologers' claims by saying that all this is very doubtful, which it is. However, we would certainly have to admit that the empirical evidence in support of astrology is greater than that for the existence of superstrings. Studies have been carried out purporting to show that it is possible to observe correlations between the positions of the planets at the moment of an individual's birth and the occupation that they later adopt. I hasten to point out that the observed correlations are small, that the studies are controversial, and that the results don't seem to conform to standard astrological ideas. Moreover, a very careful and meticulous study of the claims of astrology was recently carried out by Shawn Carlson, a researcher at the University of California's Lawrence Berkeley Laboratory. Carlson's study was unique in that he enlisted the help of prominent astrologers in designing it, and he made every attempt to avoid an antiastrological bias. Nonetheless, the astrologers who agreed to be tested—supposed to be among the best in the country—performed at a level no better than chance.

Although astrology didn't fare very well in Carlson's study, its claims can at least be tested, while those of the superstring theorists cannot. It appears that if we define "science" as something that can be subjected to experimental test, then astrology is scientific, while superstring theory is not.

Clearly, something is wrong here. We must at least admit that the distinction between science and pseudoscience isn't as simple as we like to think.

This question has been discussed at length by many philosophers of science. According to Karl Popper, a good scientific theory is one that withstands repeated efforts to disprove it, while pseudoscience depends on ideas that are not falsifiable. Thus, Einstein's theories of special and general relativity are scientific because they suggest experimental tests that can be carried out to prove or disprove them. Astrology, according to Popper, isn't a science because astrologers can always explain erroneous predictions away—if that conjunction of Jupiter and Saturn didn't have its predicted effect, an astrologer can always say that its influence was negated by the things Mars and Venus were doing.

Popper's work has undoubtedly contributed significantly to our understanding of science—he must be considered the most important philosopher of science of the twentieth century. Nonetheless, his criterion for distinguishing between science and pseudoscience doesn't always work. In particular, its applicability to some current work in physics (where lately, theory has shown a tendency to outrun experiment) is uncertain. The superstring theorists, moreover, are not the only ones inventing ideas that can't be tested in the laboratory.

A number of other philosophers have attempted to give accounts of scientific method that would allow us to distinguish science from pseudoscience. According to Thomas Kuhn, author of the extremely influential *The Structure of Scientific Revolutions,* science is a puzzle-solving activity, while pseudoscience isn't. Scientists attempt to find the answers to problems, while the practitioners of pseudoscience don't. Kuhn's criterion does seem to work in the case of astrology, but it fails elsewhere. For example, no one would deny that UFO investigators are trying to solve puzzles; we simply think that they see puzzles where none exist, or that they have the wrong answers.

There may be no simple way to make the distinction between science and pseudoscience; no philosopher has yet suggested a method impervious to criticism. Perhaps we must employ several criteria simultaneously.

⊂⊐

Before I go on, it might be well if I inserted a disclaimer. I am not a philosopher of science, and I have no illusions that I could succeed where such eminent writers as Popper and Kuhn have failed. When I speak of telling what is science from what isn't, my aims are somewhat more modest. I don't seek to invent any abstract account of scientific method that will reveal science's true character. All I want to do is look at science and pseudoscience in the manner of an anthropologist, and see what can be observed.

So far, the only science I've discussed is superstring theory. I wanted to say something about it because it lies on the frontiers of theoretical physics and therefore has a certain intrinsic interest. However, examining this theory in detail would probably not be very productive, since no one knows whether or not it's correct. Perhaps it would be better to examine a scientific topic old enough to be reasonably familiar to most people: Galileo's advocacy of Copernicus's idea that the sun, and not the earth, is the center of the solar system, that the earth revolves around the sun, and is not the motionless center of the universe.

As we all know, Galileo was silenced by the church; he was threatened with torture and forced to recant. Although no one would dispute that this was a major setback for science, it isn't possible to maintain that Galileo was entirely right and his persecutors entirely wrong. It is even possible to defend what they did, for the church had given Galileo permission to teach the Copernican system as a hypothesis that was useful for making astronomical calculations. His "crime" consisted in going further and arguing that the theory was more than a useful aid to calculation, that it had to be regarded as true.

Among the interesting aspects of the case is that Galileo did not have a shred of evidence to indicate that the earth revolved around the sun. Not until 1729—eighty-seven years after Galileo's death—did scientists devise an experiment proving that the earth really was in motion. They found that stars seemed to shift their positions slightly when the earth was on opposite sides of the sun, and concluded that the effect must be caused by the earth's revolving around the sun. Furthermore, an experimental demonstration of the idea that the sun and stars seem to move because the earth spins on its axis wasn't obtained until 1851, when the French scientist Jean Foucault observed that a large, heavy pendulum would begin to swing in a different direction as the hours passed. The deviation in the pendulum's motion was slight, but it was observable, and could mean only that the earth was twisting under it.

Galileo did construct a telescope, with which he made several important discoveries: He saw that the planet Venus had phases, like the moon, and he discovered four of the moons of Jupiter. Naturally, Galileo attempted to make much of these facts, yet such observations did not prove very much. At best, they showed that there was more to the cosmos than had previously been believed, for the old Ptolemaic conception of the universe had no place for moons around Jupiter. Yet the existence of these moons hardly demonstrated that the earth went around the sun.

At the time, no one could even be sure that the things Galileo saw through his telescope weren't illusions. The telescope was a very recent invention, and no one understood completely how it worked. An accurate law of the refraction of light (the bending of a ray of light when it passes through an object like a glass lens) wasn't discovered until twelve years after Galileo began his astronomical observations. Furthermore, the lenses from which early telescopes were made weren't of particularly good quality, certainly not by modern standards. The critics who refused to look through Galileo's telescope, saying that it produced illusions, made an argument that could easily have turned out to be valid. In that day, illusions were sometimes present in such instruments. For that matter, it wasn't even certain that some of the objects Galileo saw existed anywhere but in his imagination—on more than one occasion, the people Galileo had persuaded to look through his telescope had been able to make out nothing.

At best, Galileo had nothing but circumstantial evidence for the validity of Copernicus's theory, but, since he had committed himself so strongly to the idea that the earth moved around the sun, he needed a clinching argument of some kind. Unfortunately, he didn't have one. So, in the end, he fell back on what can be described only as a crackpot theory about the ocean tides. When Galileo published his *Dialogue on the Great World Systems* in 1632, the German astronomer Johannes Kepler had already formulated a correct theory of the tides. Kepler had hypothesized that the ocean levels rose and fell because the moon exerted a pull on the water in the seas. But Galileo would have nothing to do with Kepler's theory. Dismissing it as astrological nonsense, he invented a crazy hypothesis of his own.

The tides, Galileo said, came about because the water in the ocean basins sloshed back and forth as the earth moved.

The only problem with this theory was that it was nonsense. And Galileo should have known it was nonsense, because it contradicted his own correct theories about inertia. He had pointed out previously that the atmosphere is carried along with the earth as it moves; we do not experience strong winds in a direction opposite to that of the earth's rotation. Similarly, when a heavy object is dropped from a tower, it falls vertically to the ground. It participates in the earth's motion; it does not fall behind the rotating earth as it descends. To believe that the water in the oceans should behave differently was absurd; it, too, should be carried along as the earth spun on its axis.

When I dwell on Galileo's mistake, I am not attempting to belittle him. On the contrary, his insistence that the theory of Copernicus *had* to be true, despite his lack of any real evidence for it, is a mark of his genius. If, in the end, he fell back on a crank theory to clinch the argument, we have to forgive him for his lapse; for he did so only because he advocated the heliocentric theory with such passion.

We should not conclude that because he lacked evidence for his ideas, Galileo was not really a scientist, or that the hypothesis he advocated was not really scientific. Galileo's contemporary Kepler and, later, Isaac Newton, behaved no differently. The experimental proof of the motion of the earth around the sun was obtained two years after Newton died. Yet, by the time of Newton's death, the heliocentric theory had gained almost universal acceptance. Newton's law of gravitation had explained the workings of the solar system so well that no one worried very much about the lack of empirical evidence.

By contrast, theories that are not the least bit scientific sometimes do seem to have empirical confirmation. For example, the theory Emmanuel Velikovsky expounds in *Worlds in Collision* implies that the planet Venus should have a very high surface temperature. In 1950, when Velikovsky's book was published, scientists generally believed that Venus was relatively cool; during the late 1950s, they began to accumulate evidence to indicate that the planet might be much warmer than they had believed. Finally, when the Soviet space vehicles entered the atmosphere of Venus during the late 1960s, these data were confirmed: The temperature of Venus turned out to be approximately 460 degrees Celsius (about 860 degrees Fahrenheit). When scientists were confronted with this evidence, they did not, of course, decide that Velikovsky's theory was correct—they knew it was a crackpot theory before the data about Venus were obtained, and they knew it was a crackpot theory after Velikovsky's prediction turned out to be correct. According to Velikovsky, Jupiter ejected Venus as a comet thousands of years ago, which then passed close to the sun. In the process, it was heated to incandescence, and has not had enough time to radiate away all its heat. Therefore, Velikovsky concluded, Venus must still be very hot today—which it is.

This is not, however, a very good empirical confirmation of the theory. Scientific theories are generally expected to give quantitative predictions, but Velikovsky's does not. It says that Venus should be hot, but doesn't say how hot. Furthermore, the theory implies that Venus should be gradually cooling off, but observations made over a period of years show that this is not happening. Finally, some of Velikovsky's other predictions turn out to be false. He says, for example, that the clouds of Venus are made of hydrocarbons (chemical compounds of hydrogen and carbon; petroleum is a mixture of hydrocarbons, for example). However, in 1973, it was established that these clouds were composed primarily of sulfuric acid vapor. The atmosphere of Venus turns out to be approximately 93 percent carbon dioxide. The other seven percent is mostly nitrogen, water vapor, and carbon monoxide. Traces of other gases are present, but hydrocarbons are not.

Nonetheless, there seems to be more evidence in support of Velikovsky's theory than there was for Copernicus's in Galileo's time. Some of Velikovsky's predictions have turned out to be wrong; in Galileo's time, Copernicus's hypothesis seemed to be wrong in some respects. For example, it was known that, if the earth moved, the stars should seem to shift their positions as it did so. This effect was not observed. It wasn't until much later that scientists realized that the shift did take place, but was not easily seen because the stars were much farther away than anyone had thought possible.

It seems that, if we want to call Galileo a scientist, and maintain that *Worlds in Collision* is pseudoscience, we had better find some better arguments than those uncovered so far. We can't say that one theory was bolstered by experimental evidence, while the other was not: There was too little evidence in support of Galileo's position. Nor is it obvious that any of the proposed philosophical criteria work well, either. For example, Velikovsky's theory satisfies Popper's criterion of experimental testability. When we examine Velikovsky's theory in detail, however, its pseudoscientific character emerges more clearly. Velikovsky claimed that Venus had been ejected from the planet Jupiter thousands of years ago based on his interpretation of various ancient myths, which he believed to describe real historical events. According to a Greek myth, the goddess Athene was born from the brow of Zeus; Velikovsky took this to be an allegory of the birth of Venus from Jupiter. (The birth was accomplished by a kind of cesarean section: Hermes, who, among other things, was a physician, split open Zeus's skull with a wedge, whereupon Athene sprang forth, clad in full armor.) Zeus is identified with the planet Jupiter (the Roman name for Zeus). However Aphrodite, not Athene, is the goddess ordinarily identified with Venus. Velikovsky claimed there were reasons for believing that this identification was incorrect. He said Athene really represented Venus, while Aphrodite was originally a symbol of the moon.

As I have said, Velikovsky claims that Venus originated as a comet expelled from Jupiter around the middle of the second millennium B.C. Venus then followed an erratic path around the solar system, passed very near to the earth on several occasions, and eventually collided with Mars. This collision forced Venus into the stable, nearly circular orbit it occupies today. The passages of Venus near the earth are supposed to have been the cause of certain catastrophic events described in the Old Testament and by various ancient myths. In Velikovsky's view Venus was responsible for the plagues that Jehovah visited upon the Egyptians when the pharaoh would not allow the Israelites to leave Egypt. Venus caused the parting of the Red Sea and produced the manna that fell in the wilderness. A later collision between Venus and the earth stopped our planet's rotation entirely and caused the sun to stand still for

Joshua. After Joshua had won his battle, Venus conveniently set the earth spinning again.

It is easy to see why scientists are reluctant to call this sort of thing science—planets don't go bounding around the solar system like this. No one has ever observed any body—large or small—being ejected from Jupiter, and there is no reason whatsoever to think that such an event should have taken place several thousand years ago. Moreover, Jupiter is the most massive planet in the solar system, with a gravitational field much stronger than any other planet's. An enormous quantity of energy (roughly equal to that radiated by the sun in a year) would be required to eject a body the size of Venus from it. Where might this energy come from? Velikovsky doesn't say. Nor does he explain why Venus didn't burst into fragments when this violent event took place.

Furthermore, according to the theory, Venus must have remained in the earth's vicinity for a period of months when the two planets first collided—a certain amount of time had to pass between the first of the biblical plagues and the falling of the manna in the wilderness. But what forces could have kept the planets together for this length of time and then allowed Venus to fly off into space? Velikovsky does not explain this, either. For that matter, how did the earth and Venus survive the encounter at all? They must have been very close to one another, close enough for their atmospheres to intermingle. According to Velikovsky, the biblical plague of flies and the manna in the wilderness both came from Venus, which could not have happened if Venus was very far away. However, if the planets did approach one another that closely, tidal forces would have caused both bodies to burst into fragments. Velikovsky claims that colliding planets do not destroy one another because the impact is cushioned by the planets' magnetic fields. However, the earth's magnetic field is very weak, and, as far as we can tell, Venus has no magnetic field at all. The earth's magnetic field is so weak that it can't move even a small object like a pin or a needle. It can move a compass, but only because the compass needle is magnetized itself and delicately balanced as well. The magnetic field of Venus, if any exists, could be no more than one ten-thousandth of the earth's, according to measurements made by a Soviet space vehicle in 1967.

Obviously such magnetic fields could not cushion the impacts between planets.

Velikovsky's theory is too farfetched to be credible. The ideas upon which it is based seem fantastic, because accepting them would mean ignoring numerous well-established principles in physics, chemistry, astronomy, celestial mechanics, and biology. The theory depends upon the existence of unknown sources of energy, on tidal forces that fail to act, on magnetic fields that become mysteriously strong, and on unknown forces that keep planets together for months and then allow them to drift apart. The theory ignores the fact that, to stop the earth's rotation, it would be necessary to expend so much energy that the planet would melt. It assumes that flies can somehow evolve and live in the atmosphere of Venus. It requires us to believe that the hydrocarbons Velikovsky says exist on Venus were somehow converted into the carbohydrates that presumably fell as manna (or perhaps Velikovsky simply did not know the difference between hydrocarbons and carbohydrates). It's just too bizarre and complicated to be true.

▭

"It's just too bizarre and complicated to be true." That statement seems to be nothing more than the gut reaction of a scientist when confronted with ideas like this. Yet it may also contain a clue to understanding the difference between science and pseudoscience.

Consider this: If Galileo didn't have any real evidence for Copernicus's heliocentric hypothesis, how could he have been so certain that it was correct and that its predecessor, the Ptolemaic theory (named after the second-century Greek astronomer Ptolemy), was wrong? Obviously, he realized that Ptolemy's theory was just too bizarre and complicated to be true. In the Copernican theory, the planets—including the earth—move around the sun in circular orbits. (Kepler was later to replace the circles with ellipses; however, most planetary orbits are nearly circular.) In Ptolemy's geocentric theory, on the other hand, it was necessary to introduce all sorts of complicated mathematical devices to align theory with observation. The orbits of the planets contained epicycles, for example. An epicycle can be thought of as an orbit within an orbit; in Ptolemy's theory, in other words, there were wheels within wheels. Even when this modification was made, things still wouldn't come out right, and the more accurate astronomical observations became, the more difficult the theory was to sustain. By Galileo's time, astronomers had introduced a number of additional mathematical devices, known as eccentrics, equants, and deferents. So many gears and wheels were added that it was impossible to believe that anything like this could exist in nature. The theory was certainly a monument to human ingenuity, but it couldn't represent anything God had created, and most of Galileo's learned contemporaries despaired of finding any theory that could correctly describe the workings of the solar system. In the eyes of many, such a thing was not possible. The best that could be hoped for was to "save the appearances" by assembling a set of mathematical devices that would allow the prediction of planetary motions. In other words, astronomers had generally given up hope of understanding the movements of the planets.

The Copernican system, on the other hand, was clear and logical. It became even clearer and more logical when Kepler worked out the laws of planetary motion, and when Newton explained these motions with a simple law of gravitation. By the time of Newton, talk of "saving the appearances" had long ceased. Science had a theory that worked. It worked so well that finding an experimental proof of the earth's motion was almost an afterthought.

Such a process is fairly typical in the history of science. Contrary to what many people think, the chief characteristic of science is not the patient accumulation of data. Science is a way of understanding the world, which, as much as any of the arts, depends upon the creative intellect. Galileo became a great scientist not because he was a better or more patient observer than his contemporaries, but because he had more imagination than they did. As we all know, though, there is such a thing as a disciplined imagination—and such a thing as an undisciplined imagination as well. I think we may safely assume that Galileo possessed the former, and Velikovsky the latter. A skeptic might ask whether I seriously mean that such a criterion can be used to distinguish between correct and incorrect theories. Do I really claim that if a theory sounds crazy, then it probably is crazy, and that if it appears clear and logical, it probably is correct? Not quite. I'm not saying anything about how we distinguish a correct theory from an incorrect one. I'm speaking only about the difference between scientific and unscientific ideas. Obviously, a clear and logical theory can turn out to be wrong.

Furthermore, correct theories sometimes seem crazy when they are first propounded. Some of Einstein's ideas certainly sounded bizarre at first. According to an often repeated story, the great Danish physicist Niels Bohr once objected to a new theory on the grounds that it was "not crazy enough." If a theory is crazy, or unorthodox, or seemingly bizarre, that does not make it pseudoscientific. Crackpot and pseudoscientific theories are bizarre in a particular way. They tend to ignore long-established scientific ideas. They operate in a world of their own, not in the world of scientific discourse.

There is nothing wrong with maintaining that an established theory is incorrect. Einstein did that; so did the physicists who developed quantum mechanics. But if scientists want to overturn accepted ideas, they have a responsibility to explain why these ideas are erroneous. They can't simply behave as though centuries of accumulated scientific knowledge don't exist. But this is precisely what Velikovsky does. If we wanted to accept the ideas in *Worlds in Collision,* we would have to throw out everything we have learned about celestial mechanics since the time of Newton, without even knowing why. As for astrology, although astrologers don't say that any accepted ideas in physics are wrong—in fact, they don't seem to have any theory at all—they claim that certain observed correlations exist between planetary configurations and people's lives. The reasons why these correlations exist presumably are unknown. If astrology is correct, then either unknown forces allow the planets to influence our lives, or strange, noncausal connections of some kind operate in the universe. The latter idea is becoming quite fashionable these days, at least in so-called New Age circles. Yet, it is just as bizarre as anything in *Worlds in Collision.* If these noncausal connections really exist, then half of what we know about physics is nonsense.

We might also consider the case of parapsychology. Most physicists and most psychologists tend to think that the evidence for the existence of such phenomena as telepathy, clairvoyance, precognition, and psychokinesis is dubious at best. I suspect, however, that this isn't the real reason for their skepticism, since scientific theories, including Copernicus's, have been accepted on less evidence. The reason that scientists are skeptical about the existence of ESP is that the idea is crazy. Parapsychologists have been able to suggest no reasonable mechanism for ESP. This is what has created the skepticism, rather than unconvincing results or the fact that both subjects and prominent experimenters have been caught cheating.

If we wanted to accept such a thing as telepathy, for example, we would have to conclude that our minds can communicate with one another in some spooky, immaterial way. Accepting the existence of other kinds of psychic phenomena is even worse. An acceptance of psychokinesis entails a belief that the mind can cause certain physical laws, such as the law of conservation of momentum, to be violated. To believe in precognition, we would have to accept that the past can sometimes be influenced by the future.

I do not say that such things are impossible. For all I know, the law of conservation of momentum does have loopholes, and our ideas about causality are not always strictly correct. I am only saying that, if a theory is to be considered scientific, it must show how these violations can come about. A theory that doesn't show this can be viewed as philosophy or mysticism, but it certainly isn't science. In other words, if the existence of psychic phenomena is ever accepted, it will not be because someone accumulates experimental evidence that is better than that which exists today, but because someone comes forward with an explanation that makes the existence of such phenomena seem reasonable. Similarly, scientists will never accept that it is even remotely possible that astrology is valid until someone tells them how these correlations with planetary configurations come about.

People have been arguing about astrology since ancient times. Scientists have been expressing criticisms of Velikovsky's theory since *Worlds in Collision* was published in 1950. Perhaps it might be worthwhile to examine a more recent pseudoscientific theory, one that has received a great deal of attention recently, especially in New Age circles: the hypothesis of formative causation put forward by the British plant physiologist Rupert Sheldrake.

When I say that Sheldrake is a crackpot, or that his theory is not scientific, people sometimes say to me, "But he's a scientist, isn't he?" Or they ask, "But doesn't he suggest ways that his theory can be tested?" I have to answer that he is a scientist and he does suggest experimental tests. Nevertheless, I continue to maintain that his ideas are not scientific. Sheldrake's hypothesis is very simple. He claims that things have the form they do not because they are shaped by physical laws, but because other things had similar forms in the past. Biological form is nothing more than a kind of habit. It has nothing to do with genes or DNA. For example, according to Sheldrake, the first field mouse (like the first anything else) came about more or less by accident. Once that mouse came into being, a "morphogenetic field" was created that intended to impose mouse-like forms on animals born thereafter. As more mice were created, these fields became stronger, and it became increasingly likely that the mice's offspring would also be mice. If the theory is correct, the fact that we look like human beings, and not like bears or turtles, can be attributed wholly to chance. Supposedly, some of our ancestors took on human, or humanlike, form more or less by accident. Once this form existed, it was more likely to arise in the future. By now, so many people have assumed this form that it is practically impossible for our offspring to be anything else.

The theory has applications outside the field of biology as well. Indeed, it has a characteristic that we often encounter in crank theories: It explains everything. For example, it can be applied to behavior. Thus, if I were to teach some rats to run through a maze, their maze-learning behavior would create morphogenetic fields that would make it easier for other rats to find their way through similar mazes. It wouldn't make any difference if these other rats had never had any contact with mine. If I make some rats run a maze in San Francisco this week, their experience will affect the behavior of rats in Tokyo next year, even if the Japanese experimenters are not aware that I or my rats exist.

Sheldrake's theory purports also to explain phenomena that are observed in the fields of physics and chemistry. Thus, crystals have the form they do because similar crystals existed in the past. According to the theory, once a chemical has crystallized in a particular way, it can more easily crystallize in the future. The fact that snowflakes are made of hexagonal crystals, then, has nothing to do with the laws of physics. Rather, long ago, some primeval snowflakes formed hexagonal crystals by chance. Atoms and molecules, too, are supposedly shaped by morphogenetic fields. The electrical attraction that binds a positively charged atomic nucleus to its negatively charged electrons is an aspect of a morphogenetic field. These fields also ensure that water is water, that oxygen is oxygen, and even that electrons are electrons.

Morphogenetic fields can't be detected, of course. They have neither mass nor energy. We have no reason to expect them to obey any of the physical laws that have been found to describe the movement of particles or the propagation of waves, or to be attenuated as they move through space or propagate over time.

Scientists often say that a good theory is parsimonious. The best theories are said to be those that explain the most with a minimum of assumptions. According to this criterion, Sheldrake's theory is marvelous indeed, since it tells us that everything we observe in the universe around us can be explained in terms of fields that can be neither seen nor felt. However, if something like this turned out to be true, then most of what we know about physics, chemistry, molecular genetics, and biology would suddenly become irrelevant. Perhaps this is why it is difficult to consider such theories as "science." In order to accept them, we would have to give up too much that is simply too clear, logical, and well confirmed. Much of the knowledge we've accumulated over the last several centuries would suddenly become meaningless.

Many people think that scientists denigrate the theories of people like Velikovsky and Sheldrake because those theories contradict orthodox science, which indeed they do. Scientists generally demand that a new theory must show why the old one is unreasonable, or must demonstrate that it is a more general version of the old one. When Einstein propounded his theories of relativity, he didn't overthrow any of Newton's laws of gravitation or of motion. On the contrary, he explained why Newton's theories worked so well. He demonstrated that the equations of relativity reduced to Newton's laws when velocities were not very large, or when gravitational fields were not very intense.

I don't claim that Sheldrake's ideas are necessarily wrong, only that they are not scientific. The great majority of pseudoscientific theories do of course turn out to be incorrect. It is possible to imagine that almost anything is true. We can imagine mountains of pure gold on the moon if we want to. We can imagine animal shapes in clouds, and faces on the planet Mars. We can imagine that the ancient Israelites ate manna that fell from Venus. The mere fact that we can imagine something doesn't imply that it is the truth. That is why the English language contains such words as "fantasy" and "reality."

Incidentally, Sheldrake's theory of formative causation recently did fail an experimental test, when scientists used a computer to synthesize a new kind of silicon "crystal" and observed that the crystal did not become easier to synthesize thereafter. I can't say that I was very surprised by the result; I was more surprised that the experiment was even done, since most scientists would not consider Sheldrake's theory worth testing.

During the last few years, I've found myself becoming rather pre-occupied with pseudoscience. I don't think this is the result of a fascination with oddball ideas, however. I simply seem to have realized that one can sometimes gain insights into what something is by looking at what it is not. An understanding of nonscience tells us something about science.

Science isn't wild speculation. If it were, we would have to call Velikovsky and Sheldrake's notions "scientific." Science isn't merely the patient accumulation of data. Many people still believe the description of scientific method Francis Bacon proposed around the beginning of the seventeenth century to be basically correct. According to Bacon, scientific laws are supposed to be generalizations from observed facts: To do science, one must accumulate some data, ponder the results, and then deduce a theory from them. Yet science doesn't work this way, and never has. The construction of a scientific theory is an act of creative imagination. Scientific innovation comes about when a scientist is confronted by something—usually a baffling theoretical problem—that he can't seem to understand. He looks at it, and looks at it, and feels completely baffled. If he is lucky, he may have a sudden flash of insight that causes everything to make sense. When this happens, a new theory has come into being.

After a theory is born, scientists begin to perform the experiments that will tell them whether or not it is correct; as Popper would say, they attempt to falsify it. Theories are not generalizations from experiment; on the contrary, it is theory that tells scientists what experiments to perform. Sometimes a theory is so compelling— sometimes it seems to explain accumulated data so well—that it is accepted before the crucial experiments can be carried out. That is why Copernicus's heliocentric hypothesis gained such wide acceptance long before anyone figured out how to show that the earth actually moves; that is why Einstein's theory of gravitation (his general theory of relativity) became so widely admired long before much evidence existed to support it (this theory can be considered well confirmed now, but at the time of Einstein's death, in 1955, the confirming evidence was minimal).

Insights are often wrong. Numerous ingenious theories have had to be discarded because the experimental evidence didn't support them. Some theories show so much disregard for well-established ideas that we don't consider them science at all. In a way, I'm happy that pseudoscientific theories are put forward. It is sometimes necessary to expend a great deal of time and energy warning the public not to take them seriously, and it seems that no matter how much debunking scientists do, new pseudoscientific ideas continue to sprout up. However, looking at such theories in detail does throw light on the nature of the creative process—Velikovsky's and Sheldrake's theories seem to be based on insights, too. The problem is only that these ideas are developed without any regard for well-established scientific knowledge.

Pseudoscientific theories are also interesting as sociological phenomena. They often seem to support certain ideas that groups of people want desperately to believe in. Velikovsky's theory seemed to support a fundamentalist interpretation of the Old Testament. Sheldrake's hypothesis of formative causation would, if it were true, corroborate that rather vague New Age idea that everything is somehow connected with everything else. Nevertheless, when pseudoscientific theories are looked at in detail, they fail to give us any new understanding of the world; they only leave us baffled. When Velikovsky tells us that Venus was ejected from Jupiter, we scratch our heads and wonder how something so bizarre could have happened. When Sheldrake assures us that there exist undetectable fields of unexplained origin that nonetheless explain everything, we doubt his sanity.

Real scientific insight provokes a different kind of reaction. When Galileo learned of Copernican theory, he must have said to himself, "Yes, that explains everything. Yes, that's the way it must be. The old system was too cumbersome, too complicated." Similarly, when Einstein proposed his special theory of relativity in 1905, some scientists were skeptical at first, but even they eventually realized that Einstein had found the solutions to problems that had been puzzling physicists for decades, such as the origin of the sun's energy. As an innovative new *scientific* theory gains acceptance, our understanding of previously accumulated facts becomes more penetrating, more coherent, somehow larger.

Small Science

JAMES E. LOVELOCK

. . . in recent years, the "purity" of science has been ever more closely guarded by a self-imposed inquisition called the peer review. This well-meaning but narrow-minded nanny of an institution makes scientists work according to conventional wisdom and not as curiosity or inspiration moves them. Like the inquisition of the medieval church, it has teeth and can wreck a career by refusing funds for research or by censoring publications. It is not unusual for scientists to be so constrained as to exhibit a finicky gentility or become, like medieval theologians, the captives of dogma. But to those aware of their plight I say: "Soon will the bonds of the peer review system be rent asunder. Come join me in independence. You have nothing to lose but your grants."

For the past twenty-five years I have tried to do science in the way recommended by Fritz Schumacher. I take his advice to think globally by acting locally most seriously. I practice global science as a family business done at home, not as some vast, remote, and potentially dangerous activity.

I live with my family in a mill cottage on the River Carey in West Devon near the border with Cornwall. My laboratory is a room built on to our cottage. The nearest house is half a mile away. The village of St. Giles on the Heath is two miles away.

I have always thought that science was something to be done at home, like writing, painting, or composing music. After all, there is nothing unusual about an artist's doing creative work at home. Indeed, the very idea of an artist's painting portraits in the department of fine arts of some university, or of a novelist's traveling daily to the Institute for Creative Writing, seems absurd. So why shouldn't science be done at home?

I don't mean to be prescriptive, and I recognize that much of science needs teamwork and a large scale of operations, just as medieval cathedrals required large teams of artists and craftspeople to build them. It is just that science in modern times has no place for the individual thinker or experimeter. I will try to show that this lack of individuals with the time and opportunity to wonder about and explore the world has grievously weakened our understanding of the natural environment.

Part of the problem arises because most of what people call, or think of as, science is, in fact, technology. A scientist is someone with the time and inclination to wonder, and who then expresses a personal view of the natural world as theories, or ideas, that later can be tested by the accuracy of their predictions. There are very few of these, and most scientists are like the creative and highly talented individuals who write the jingles and make the glossy pictures that advertise commercial products. They would not claim that their work was art, still less that they were free to write or paint as the spirit moved them. Their position is not different from that of most scientists in the world today.

You might think that the tenured academic scientist working on a chosen subject in a good university is as free as an independent artist. In fact, nearly all scientists employed by large organizations, such as government departments, universities, or multinational companies, are rarely free to express their science as a personal view. We may think of them as free, but in reality they are, nearly all of them, employees; their employers may exert only gentle pressure, but that is usually all they require. Academic scientists also have to contend with an army of bureaucratic forces, from the funding agencies to the health and safety organizations. Scientists are also constrained by the tribal rules of the discipline to which they belong. A physicist would find it hard to do chemistry, and a biologist would find physics well-nigh impossible to do. To cap it all, in recent years, the "purity" of science has been ever more closely guarded by a self-imposed inquisition called the peer review. This well-meaning but narrow-minded nanny of an institution makes scientists work according to conventional wisdom and not as curiosity or inspiration moves them. Like the inquisition of the medieval church, it has teeth and can wreck a career by refusing funds for research or by censoring publications. It is not unusual for scientists to be so constrained as to exhibit a finicky gentility or become, like medieval theologians, the captives of dogma. But to those aware of their plight I say: "Soon will the bonds of the peer review system be rent asunder. Come join me in independence. You have nothing to lose but your grants."

My own special interest as a scientist arose from wondering about the earth and the life upon it. It developed from just wondering to become a new theory about evolution, the earth, and the life upon it. It is a theory that sees the evolution of the species of organisms and the evolution of the material earth not separately, as in the division of science into biology and geology, but as a single, closely coupled process. The self-regulation of the climate and the composition of the air, the ocean, and the rocks are seen as emergent properties arising automatically from the close coupling of living organisms and their material environment. The regulation proceeds entirely without foresight or planning. On the advice of the novelist William Golding, I called the theory Gaia, after the Greek name for the earth goddess.

Had I been a university scientist, I would have found it nearly impossible to work on Gaia. To start with, no funds would have been approved. It would have been considered much too speculative. If I had persisted and worked in my lunch hour or spare time, it would not have been long before I'd received a summons from the laboratory director. In his office I would have been warned of the dangers to my career of persisting in so unfashionable a research topic. If this had not worked, and I'd obstinately persisted, I would have been summoned a second time and warned that my work endangered the reputation of the department, and the director's own career. If I was lucky, I would have been given a leaflet on how to apply for early retirement.

The reason I chose to go freelance as a scientist was not because I was prevented from doing Gaia research—I had not even thought of it when I resigned from employment. I was employed until 1961 by the Medical Research Council at the National Institute for Medical Research, in London. No employer could have been more considerate. Conditions of work were near ideal, and I had a good salary, long holidays, and an amazing degree of intellectual freedom. Strangely, it was tenure that made me leave that comfortable nest— not the lack of it, but the difficulty of living and working with it. It was the stifling prospect of tram lines of certainty leading straight to retirement and the grave that for me killed creativity and drove me to leave. Some of us need the spur of insecurity to come alive.

In the last years of my employment at the National Institute, I had invented a number of sensitive detection devices. As Fritz Schumacher would have understood, I felt that inventions should be made in response to the needs of the world in which the inventor lives, and not in the isolation of the laboratory. Only as an independent could I go out and use my inventions to measure the things that interested me in the real world.

One invention I made some thirty years ago was a device called the electron capture detector. It is still the most sensitive, easily portable, and inexpensive analytical device in existence. It is so exquisitely sensitive that if a few liters of a certain perfluorocarbon were allowed to evaporate somewhere in Japan, we could easily detect it in the atmosphere here in Bristol a few days later, and within a year or so, anywhere in the world.

It was this device that started the environmental movement. It provided the base data that pesticides like DDT and Dieldrin were distributed throughout the global environment. It showed that they could be found in the fat of Antarctic penguins and in the milk of nursing mothers in Finland. These data enabled Rachel Carson to write her seminal book *Silent Spring* and warn the world of the ultimate consequences of the continued use of these chemicals in the unceasing battle against all forms of life that are not livestock or crops.

When I first heard that the electron capture detector was being used this way I was delighted. Rachel Carson's first interest was in the damage to wildlife and to natural ecosystems. Some parts of the chemical industry reacted in a shameful and foolish way by trying to discredit her personally. It did not work. In fact, it made Rachel Carson the first saint and martyr of the infant and innocent Green movement, and all seemed set for its development as a force to oppose the disintegration of creation by humans. Sadly and soon, the environmental agenda became concerned with narrow human interests, and with threats to our species specifically rather than with a proper concern and empathy for the natural world around us.

Environmentalism's fall from innocence was inevitable. The discovery of the ubiquitous distribution of measurable, but usually trivial, quantities of pesticides and numerous other chemicals by the electron capture detector opened a niche for what is now the environment industry. In the rich soil of this niche, pollution chemists flourished and found a new set of career opportunities measuring pesticide residues in food, water, and the air. Instrument companies grew to fill the needs of these chemists. The field was fertile also for lawyers and for regulatory agencies employing civil servants. Academic and institutional scientists found a grand new source of funding for research, and the media a whole new range of stories. The general public found in this new environmentalism a way to express its fears of things scientific and technological. People found an amazing collection of scapegoats to blame for their personal and collective ills. Not for one moment would I suggest that there was any conspiracy—there was no need for one. As the famous quatrain goes:

> You cannot hope to bribe or twist,
> Thank God, the British journalist.
> But seeing what the man will do
> Unbribed there's no occasion to.

Nothing so scares most men and women as cancer. We fear it more than any other disease and willingly donate money for research or for anything that may prevent us from dying in what we see as a painful and degrading way. The fact that the electron capture detector is specifically and uniquely sensitive to carcinogenic chemicals, and that nuclear radiation is also carcinogenic, tied environmental science to cancer. As a consequence, environmental research and action were channeled by public and political pressure to do what it was best at and what it wanted to do—namely, to measure and do research on the abundance of carcinogenic chemicals and radiation in the environment. By this means the essentially global vision of Rachel Carson, whose concern was for the natural world, was corrupted to become mainly a human interest.

I do not mean that I regard as unimportant the presence in the environment of radiation and chemicals at levels that could cause a significant increase in cancer. I mean that given the exquisite sensitivity of the instruments, it is all too easy to find carcinogens everywhere, even when the danger from them is small. As one of the German Greens said, "Chernobyl is everywhere." Of course it is. Radioactivity is so easy to measure—as are DDT, polychlorinated biphenyls, and all manner of other things.

In the evolution of environmentalism, not only was Rachel Carson's vision corrupted but also the data involved. I do not mean that false values were used; so sensitive is the electron capture detector that numbers can be given for utterly trivial quantities of pesticides and other chemicals. Before I invented the device, it would have been quite easy and reasonable to set zero as the lower permissible limit for pesticide residues in foodstuffs. In practice, zero really means the least that can be detected. After the electron capture detector appeared, zero as a limit became so low that to apply it would have caused the rejection of nearly everything that was edible. Even organically grown vegetables contain measurable levels of pesticides.

What was needed was common sense and the acceptance of the wisdom of the physician Paracelsus, who said long ago, "The poison is the dose." Even water is poisonous if too much is taken. Even one of the deadly nerve gases is harmless at the level of a picogram, easily detectable by an electron capture detector.

Perhaps some will feel that I exaggerate or give a distorted view. If so, another way to illustrate what I mean by a commonsense approach is to consider the relatively clear-cut environmental problem of ozone depletion by halocarbons. So let me relate something of my personal experiences in what has been called the Ozone War during the past two decades, and of the part played by small-scale Schumacher-type science.

In 1968 we purchased a holiday cottage in far western Ireland, on the shores of Bantry Bay. It stood on the slopes of Hungry Hill, a small mountain of warm sandstone slabs made famous in the book of that name by Daphne DuMaurier. It looked out over the broad Atlantic. Most of the time the air was so clear that islands fifty miles away could be seen from the mountaintop, but occasionally the air was so hazy that nothing more than a mile away could be seen. The haze looked and smelled to me just like the photochemical smog of Los Angeles. But how could it have reached this remote rural region? Next summer, to the disgust of my family, I took a home-made gas chromatograph equipped with an electron capture detector on holiday. I had the idea that it should be possible to decide if the haze was a natural phenomenon, or was human-made, by measuring the level of chlorofluorocarbons (CFCs), the aerosol propellant gases, in it. The CFCs are unique among chemicals in the atmosphere in being unequivocally of industrial origin. All other chemicals have natural as well as human-made sources. My idea was that if the haze was pollution, it must come from an urban industrial area and could contain more CFCs than does clean Atlantic air. On the first few days of our holiday it was sparkling clear, and I was surprised when I measured the air to find a strong signal for one of the CFC gases at about fifty parts per trillion. A few days later, the wind shifted, and an easterly drift of air blew from Europe. With it came the haze and the pleasant confirmation of my idea about the origin of the smog, for the hazy air contained three times as many CFCs as the clear air. The haze *was* human-made. Later investigations showed it to be photochemical smog, rich in ozone, and to have come from southern France and Italy, having drifted in the wind nearly a thousand miles, carrying the exhaust fumes of the millions of cars of European holiday makers.

There this small investigation might have ended. But being curious and having no employer to tell me what I should be doing, I wondered about the fifty parts per trillion of CFCs in the clean Atlantic air. Had it drifted across the Atlantic from America? Or, more exciting, were the CFCs accumulating in the earth's atmosphere without any means for their removal? To find out, the only thing to do would be to travel by ship to the Southern Hemisphere and back, and measure the CFCs as the ship traveled across the world. On my return to England, I applied to NERC and SRC for funds to make a more robust apparatus and for the expedition to the Southern Hemisphere and back. The idea of an individual scientist's attempting such a grandiose measurement was more than the peer reviewers, used to funding large and expensive research, could stomach. My two applications were rejected. The major part of the cost of the expedition was funded from our housekeeping money, and the acknowledgment at the end of my paper was to my wife, Helen Mary, for her generosity in permitting its use in this way. I am also grateful to the civil servants of NERC who let me travel on one of their ships, the *Shackelton,* to Antarctica and back. I was able with my homemade apparatus on board to measure the CFCs and other gases across the hemispheres.

This journey of research revealed not only the presence of the CFCs but also the quite unexpected and ubiquitous distribution of other gases: methyl iodide, carbon tetrachloride, and the sulfur gases dimethyl sulfide and carbon disulfide. These, it now seems, may be even more significant a discovery than that of the CFCs. It was all done as small science; the apparatus was so simple, I was able to make it in a few days. It ran without failure throughout the six-month voyage. Each day, measurements were made in turn by me, my colleague Robert Maggs, and one of the ship's scientists, Roger Wade. The total cost of the research, including the apparatus, was a few hundred pounds.

The results of the voyage were published in three articles in *Nature*. The discovery of the global distribution of the CFCs and the sulfur and iodine gases can be credited to Schumacher-style science. It would never have been made by big science, for, as the peer reviewers said in their rejections of the grant applications, the measurement was pointless and in any event could never be made because there was no instrument sensitive enough for their detection.

The findings that the CFCs were accumulating in the atmosphere almost without loss stimulated Sherry Rowland and Mario Molina to write their historic paper in *Nature*, warning that the slow breakdown of the CFCs in the stratosphere would release chlorine, and that this could catalyze the destruction of stratospheric ozone and hence increase the flux of ultraviolet radiation at the earth's surface.

When Rowland and Molina made public their concern about the potential of the CFCs to deplete stratospheric ozone, I was very skeptical. Not that I doubted the excellence of their science or the validity of their hypothesis. What I did doubt was that the fifty parts per trillion of F11 and the eighty parts per trillion of F12 in the air in the mid-1970s were, at that time, a significant threat. I also made public my view that the potent greenhouse properties of these compounds might be a more serious danger, when and if they increased to the levels now present. These seemed to me to be things to watch, and curtail if and when they became a menace. Hardly something to lose sleep over or legislate against in the 1970s.

But once it was realized that yet another group of industrial chemicals, no matter how harmless and useful they were in the home, might indirectly cause cancer through ozone depletion, there was an outburst, an explosion, of hype. Newspapers carried banner headlines saying, "These Chemicals Will Destroy All Life on Earth." As a result, funds flowed as never before; something near to 100 million pounds must by now have been spent by big science on stratospheric research connected with the CFC ozone affair. This time, perhaps having learned its lesson, the chemical industry was amazingly cool. Through its representative body, the Chemical Manufacturers Association, it actively supported the work of the scientists—even those whose findings were in favor of a cessation of the production of CFCs. This public-spirited behavior deserves recognition. Cynics have claimed that the generous and open policy of the industry was deliberate and in the hope that the rival scientific groups would neutralize one another; even if true, it seems an enlightened approach compared with the hostility and disinformation that met Rachel Carson's book. In their book *The Ozone War*, Harold Schiff and Lydia Dotto describe a battle not between industry and environmentalists, but between rival groups of scientists for the right to defend that sacred place in the sky.

The public and the Green movement have been misled into thinking that this is the most serious of environmental concerns. It is in fact extremely interesting to scientists, but it is only a part of the environmental threat that looms ahead.

Nearly twenty years have passed since the prevalence of CFCs in the air was discovered. In that time they have increased in abundance by 500 percent and they are still growing at 6 percent per year. I have no doubt that the release of these gases should be stopped, and stopped immediately. Not because of their ozone-depleting tendency, but because stopping the emission of CFCs is the one positive thing we can do immediately about the menacing greenhouse gas accumulation. I accept that ozone depletion could become a problem, but ultraviolet is more dangerous to people, especially white-skinned people, than it is to the planet.

What about the ozone hole? Is this a serious menace to life on earth? Will it spread? The first point to note about the ozone hole is that it was discovered by a lone pair of British observers, Farmer and Gardiner, using an old-fashioned and inexpensive instrument on an expedition in the Antarctic, just as the presence of CFCs was discovered seventeen years ago. It is true that big science, as a consequence of the ozone war, has been able vastly to improve our understanding of the complexities of the atmosphere, but it is a scandal that the vast sums spent on expensive computer models of the stratosphere, and on the big science of satellite, balloon, and aircraft measurement, failed to predict or find the ozone hole. Worse than this, so sure were the computer modelers that they knew all that mattered about the stratosphere, that they programmed the instruments aboard the satellite that observed atmospheric ozone from above to reject data that were substantially different from the model predictions. The instruments saw the hole, but those in charge of the experiment ignored it, saying in effect, "Don't bother us with facts. Our model knows best." The ozone war is littered with stories of this kind of military incompetence. I well recall being told in Washington that my original measurements were erroneous because they did not fit the model of ozone depletion. Only three years ago also in Washington I was told that the instruments measuring ultraviolet radiation at the earth's surface must be faulty because they all showed a decrease in ultraviolet and therefore less danger of skin cancer. My response to the scientists there was that if that was how they reacted to the message from their instruments, I would not like to be a passenger aboard an aircraft of which they were the pilots. Subsequently, a paper has been published in *Science* confirming that the instruments were right and that ultraviolet radiation has *decreased* in intensity continuously throughout the last ten years of measurement in the continental United States.

Second, the ozone hole is at least as much a consequence of methane gas as it is of CFCs. Now, methane, or natural gas, is mainly the product of agriculture and forestry in all parts of the world, although some does leak from natural-gas installations and pipelines. Methane is not often mentioned as a pollutant, yet it is probably the most dangerous substance that we are injecting into the atmosphere. Methane is a key agent in the ozone-hole phenomenon; but, much more seriously, it is a greenhouse gas that before long may overtake carbon dioxide in significance. The production of CFCs is already curtailed and soon, I hope, will be banned altogether. Carbon dioxide can be cut back, if we have the will. But to stop the excessive release of methane gas from rice paddies and cattle farts is all but impossible.

I hope that the Green movement may learn something useful from this account of the ozone problem as seen from the perspective of small science. Environmental groups are often hostile to science and to scientists, and, as I have tried to show, they are right to be skeptical about the claims of big science. Yet the science advisers to the Greens, who may feel objective, are in fact reflecting the needs and prejudices of the scientific community to which they belong. That community itself is divided and uncertain in the face of a world-wide decline in funds. Scientists are human and for the most part concerned with careers, pensions, security, and all the needs of modern city life. Also, it is not easy to be responsible where there is no accountability.

A consequence is that the list of priorities often given by leaders of the Greens reflects the list of working priorities of the scientific community, rather than the priorities of people—still less the priorities of the planet.

This is why until quite recently the Greens appeared to list global dangers in order of priority as follows: first, all things nuclear, whether power stations, processing sites, waste disposal, or bombs; second, ozone depletion; third, the waste products of the chemical industry. In other words, the first three priorities are things carcinogenic or mutagenic to humans. Until Prime Minister Thatcher, in a speech before the Royal Society, stole the flag by stressing the greater danger from greenhouse gases, pollutants that threaten the planet, like carbon dioxide and methane, would have been lower on the list—probably below acid rain, but higher than the destruction of the rain forests of the tropics.

As an independent scientist I see things very differently. To me the vast, urgent, and certain danger comes from the clearance of the tropical forests. Greenhouse gas accumulation may be an even greater danger in time to come, but now it is not.

The tropical rain forests are both a habitat for humans and a physiologically significant ecosystem. That habitat is being destroyed at a ruthless pace. Yet, in the "first world" we try to justify the preservation of tropical forests on the feeble grounds that they are the home of rare species of plants and animals—even of plants containing drugs that could cure cancer. They may be. They may even be slightly useful in removing carbon dioxide from the air. But they do much more. Through their capacity to evaporate vast volumes of water vapor, and of gases and particles that assist the formation of clouds, the forests serve to keep their region cool and moist by wearing a sunshade of white reflecting clouds and by bringing the rain that sustains them. Every year we burn away an area of forest equal to that of Britain, and often we replace it with crude cattle farms. Unlike farms here in the temperate regions, such farms rapidly become desert, more trees are felled, and the awful process of burning away the skin of the earth goes on. We do not seem to realize that once more than 70 to 80 percent of a tropical forest is destroyed, the remainder can no longer sustain the climate and the whole ecosystem collapses. By the year 2000, given the present rate of clearance, we shall have removed 65 percent of the tropical rain forests. After that, it will not be long before they vanish, leaving the billion poor of those regions without support in a vast global desert. This is a threat greater in scale than a major nuclear war. Imagine the human suffering, the refugees, the guilt, and the political consequences of such an event. And it will happen at a time when we in the first world are battling with the surprises and disasters of the greenhouse effect, intensified by the extra heating from the forest clearance. We may be in no position to help.

In addition to this impending catastrophe, the earth in the next few decades will pass through one of its major transitions, and we, its first social intelligent species, are privileged to be both the cause and the spectators. The event is an imminent major climate change, one that will be at least twice and possibly six times as great as the change from the last ice age until now.

Let us look back to the deepest part of the last glaciation, some tens of thousands of years ago. The glaciers reached the latitude of St. Louis in America and the Alps in Europe, the sea level was at least four hundred feet lower than now, and consequently an area of land as large as Africa was above water and covered with vegetation.

The tropics were almost as warm as now. In all, this was a rather pleasant planet to live on, and it was the home of simple natural humans just beginning to try such neat ecocidal tricks as fire-drive hunting (setting the forest on fire for a free effortless barbecue).

To understand what is before us in the next century, imagine a change in climate at least twice as great as that from the ice age until now, the start of a heat age. The temperature and the sea level will climb decade by decade, until eventually the world will become torrid, ice-free, and all but unrecognizable. "Eventually" is a long time ahead; it might never happen. What we must prepare for now are the events of the transition itself, which are just about to begin. These are likely to be surprises as unpredictable by big science models as was the ozone hole. They may be extremes, like storms of great ferocity and unexpected atmospheric events. Nature is non-linear and unpredictable, and never more so than in a period of transition.

But what of Gaia? Will she not respond and keep the status quo? Before we expect Gaia to act, we should realize that the present interglacial warm period could be regarded as a fever for Gaia and that left to herself, she would be relaxing into her normal, comfortable (for her) ice age. She may be unable to relax because we have been busy removing her skin and using it as farmland, especially the rain forests of the tropics, which otherwise are among the means for her recovery. But also we are adding a vast blanket of greenhouse gases to the already feverish patient. In these circumstances Gaia is much more likely to shudder and move over to a new stable state fit for a different and more amenable biota. It could be much hotter or much colder; but whatever it is, it is no longer the comfortable world we know.

One ray of hope is a recent discovery that also has come mainly from small science. You may remember I mentioned that one of the gases measured on the *Shackelton* voyage was dimethyl sulfide. It turns out that without the presence of this gas, there would be fewer and less dense clouds over the oceans of the world. It oxidizes in the air and produces tiny droplets of sulfuric acid, onto which the cloud droplets form. Without the white clouds over the oceans, the earth would be much hotter, for the dark ocean absorbs the sun's heat, but the white clouds reflect it back to space. Dimethyl sulfide is entirely the product of marine algae living in the ocean surface. If the output of dimethyl sulfide from the oceans increases as the world warms, the extent of heating by the gaseous greenhouse may be ameliorated.

The human and political consequences of the two geocidal acts, forest clearance and suffocation by greenhouse gases, will be the news—news that will usurp the political agenda. Soon, and suddenly, in the humid tropics, there could be a billion or more humans enduring floods and drought, perhaps with mean temperatures of 120 degrees Fahrenheit. They would be without support, in a vast arid region around the earth. All this could happen at a time when we in the North, who might otherwise come to their rescue, are facing rising sea levels and major changes in our own climate, and the most amazing surprises. These predictions are not fictional doom scenarios, but uncomfortably close to certainty. We are like a modern version of the gadarene swine, driving our polluting cars heedlessly down the slope into a sea that is rising to drown us.

I have spoken as a lone independent scientist, but I am far from being outside human concerns. We have eight grandchildren and wish to see them grow up in a world that has a future. Nevertheless, following an independent, Schumacher life-style has thrust me into the vacant position of the representative, the shop steward, for the rest of the life on the planet. My constituency is all life other than humans, and includes the bacteria and the less attractive forms of life. I have to do this because there are so many who speak for people, but so few who speak for these others, on whom the planet depends more than it does on us.

From my laboratory in West Devon I can see the stars at night and the Milky Way. In the daytime I can hear the birds and smell the earth. To see and feel the earth this way and to think of it as a living organism gives substance to the Christian concept of stewardship and turns our hearts and minds toward what should be our prime environmental concern, the care and protection of the earth itself, and especially of the tropical rain forests. It is not enough merely to be concerned for people. There is no tenure for anyone on this planet, not even a species. If we do not recognize our responsibility to our planet, we may not even reach our allotted span. So let us be moderate in our ways and aim for a world that is healthy and beautiful, and that will remain fit for our grandchildren, as well as for those of our partners in Gaia.

Chaos in Myth and Science

RALPH H. ABRAHAM

We are learning that chaos is essential to the survival of life. Our challenge now is to restore goodness to chaos and disorder, to replace Tiamat on her rightful throne, in mythology and in daily life, and to reestablish the partnership of cosmos and chaos, so necessary to creation. This will require a major modification of our mythological foundations, unchanged these past millennia: no mean feat.

Dynamical systems theory, a relatively new branch of mathematics, concerns abstract theories of motion, without reference to mass, force, or any other physical property. It contains not only chaos theory, but an extensive strategy for building models for complex processes as well. One of the most complex processes to which this strategy has been applied is the process of history, including the evolution of human consciousness.

In these new mathematical models, simple forms called attractors are observed as the representations of the stable (or observable) states. There are three types of attractors. Point attractors (also called static attractors) model the stationary states of a system; they represent the system at rest. Periodic attractors consist of a cycle of states, repeated again and again, always in the same amount of time; they represent the system in oscillation. Chaotic attractors consist of fractal (infinitely folded) sets of states, over which the model system moves, apparently at random; they represent the system in a state of agitation or turbulence.

These three mathematical objects were not discovered all at once. First the point attractor emerged, soon after Newton developed the basic mathematics for dynamical systems (the branch of mathematics now known as ordinary differential equations) circa 1700. The periodic attractor came into general use around 1850, spawning a new branch of mathematics, known as nonlinear oscillation theory. And the chaotic attractor, although known in one form (the so-called homoclinic tangle) to Poincaré in 1885, emerged into scientific consciousness only around 1961.[1]

The application of the chaotic attractor model to the various sciences began in 1974. Now known as chaos theory, it provides new methods for understanding data that previously were regarded as noise. Since 1974, a great deal of noise has been discovered to be "signal." That is, chaos theory has provided tools for analyzing chaotic data and finding order in them. One of these tools, called attractor reconstruction, actually makes a computer graphic drawing of the dynamical model (chaotic attractor) for the data.

When dynamical models change, their attractors (or observable states) may change in type or even disappear. These events, called bifurcations in dynamical systems theory, are used to model important transformations in complex systems and processes, such as cultural history. Thus, the dates 1700, 1850, and 1961 mark bifurcations in the mathematical understanding of dynamical processes. What about the ordinary understanding of these three kinds of dynamical behavior—the static, the periodic, and the chaotic?

According to a folk theory in mathematical vogue, a mathematical object will emerge into the collective consciousness soon after the human mind evolves to the necessary complexity to support the cognition of that object.[2] That is, the brain and mind evolve, consciousness evolves, language evolves, and mathematical objects are "discovered" as soon as people are able to understand them. Extrapolating this evolutionary view to the grand scale of the history of consciousness, we may transcend the level of mathematical model and look for the emergence of the kinematical concepts: static, periodic, and chaotic processes. There results a simple division of recent history into three ages, which correspond to the three kinematical concepts, or paradigmatic types of behavior: the static, the periodic, and the chaotic.

The emergence of the static concept might have preceded the development of language, but for practical purposes let us begin with this important development. No one knows whether language developed with *Homo sapiens* about fifty thousand years ago, or much earlier, or later. Since we know so little about our parent species, we will restrict our attention to the past fifty thousand years. The first bifurcation in consciousness in this hypothetical series, the emergence of the static concept into awareness, might coincide with the development of linguistic structures such as *stand still*. The first material manifestation of this idea might be the development of agriculture, the Neolithic revolution, around 8000 B.C. Thus began the Static Age.

Later came the concept of cyclic repetition. Although there are artistic representations of the phases of the moon dating as far back as 30,000 B.C., the material manifestation (or realization) of the idea is the invention of the wheel, around 4000 B.C. Here began the Periodic Age.

The chaotic process may have emerged into awareness in remote antiquity, when it was represented by symbols such as serpents, sea monsters, clouds, and goddesses, including Innana, Ishtar, and Tiamat. Perhaps because of the repression of the chaotic by the law-and-order gods of the patriarchal society, this concept did not fare well in the history of consciousness. But many concepts emerge in mathematical studies before appearing in the material record as an artifact of human society. For example, the circle concept (in the form of the cycle of phases of the moon) was known to the epi-Paleolithic mathematicians of Cro-Magnon long before its appearance as the material wheel. Similarly, the chaos concept was known as a mathematical object to Poincaré, the great French mathematician, a century ago. Yet its material realization is the computer model first observed by Ueda in November 1961. And here began the Chaotic Age.

As we now know that the solar system is chaotic, and that the motions of the sun strongly influence our global climate (especially the ice ages), we should not expect history to follow periodic cycles either. The dates of the most recent ice ages have been accurately determined, and it seems possible that Neolithic revolutions occurred in previous interglacial periods. There have been various attempts to identify the glaciations with a periodic process, but recently it has been suggested that they occur chaotically. Conceivably, a new Static Age could occur again after the next glaciation (or nuclear winter) recedes. Thus, the dynamical cycle of three ages might be recurring chaotically in the future, giving realization to ancient theories of the ages of the world. As devotees of chaos, we do not expect history to be periodic!

Cosmos means order, or arrangement, and a *cosmogony* is a theory of the origin of the universe (the root *gonos* signifying birth). Most cultures seem to have a developed cosmogony, and anthropologists see this as an indicator of cultural sophistication. In fact, the development of a cosmogony is a step in creation itself, a kind of cultural self-reference. Our own cosmogony (including the version prevalent among astrophysicists) may be understood as an outgrowth of that of the ancient Babylonians,[3] in whose early myths we find chaos playing a special role. *Chaos* now means disorder, but in ancient myths there are important variant meanings, to which we now turn.

Our literature begins in Mesopotamia, in a milieu of patriarchal dominance, at a time when the mythology of the displaced goddess culture is still in a process of transformation.[4] Some outstanding features of the widespread goddess culture of the early Neolithic are peace, partnership of women and men, and love of the earth. Also, it has been suggested, the partnership of chaos and order survived until the patriarchal domination, around 4000 B.C.[5]

The earliest written material documents the goddess worship of the Semitic people of Mesopotamia. On Ishtar, worshiped since 5000 B.C., we have the following, in the rendering of Merlin Stone:[6]

Queen of Heaven, Goddess of the Universe,
the One who walked in terrible chaos
and brought life by the law of love
and out of chaos brought us harmony
and from chaos She has led us by the hand.

In another early cosmogony, known from a Sumerian cuneiform text of circa 2000 B.C., Nammu, the goddess who gives birth to Heaven and Earth, is referred to by an ideogram signifying *sea*.[7] A contemporaneous cosmogony from Eridu, a Babylonian seaside town, is known from a bilingual (Sumerian and Babylonian) cuneiform tablet dating from before 700 B.C. and discovered in 1891. The cast of characters includes Apsu and Tiamat, the male and female aspects of the watery deep. Apsu ("the Deep") is the home of the culture-god, Ea; Tiamat, the watery chaos, is enemy of the gods of light and law, and is pictured as a dragon. The creation story begins with this description:

No holy house, no house of the gods in an holy place had as yet been
   built,
No reed had grown, no tree planted,
No bricks been made, no brick-mold formed.
No house been built, no city founded,
No city built, no man made to stand upright;
The deep was uncreated, Eridu unbuilt,
The seat of its holy house, the house of the gods, unerected:
All the earth was sea,
While within the sea was a current. . . .[8]

Later (circa 1800 B.C.) is the better known *Enuma Elish,* also known
as the *Epic of Creation.*[9] At this time Babylon had become the Big
Apple in Mesopotamia, and Bel-Merodach (better known as Marduk, the city-god of Babylon) had replaced Ea as Mr. Big. Bel was
a masculine form of Belili, the Sumerian mother goddess. Many
gods underwent sex changes during the cultural transformation (bifurcation) from partnership to patriarchal society. *Enuma Elish* is an
epic hymn in honor of Marduk and his overthrow of Tiamat and the
powers of chaos. The epic begins thus:

When above unnamed was the heaven,
And earth below by a name was uncalled,
Apsu in the beginning being their begetter,
And the flood of Tiamat the mother of them all,
Their waters were embosomed together (in one place),
But no reed had been harvested, no marsh-plant seen;
At that time the gods had not appeared, any one of them.
By no name were they called, no destiny was fixed.
Then were the gods created in the midst of heaven. . . .[10]

Then unrolls the creation scenario. The appearance of the gods of
light and order was followed by the revolt of Tiamat. Then the
forces of darkness and chaos were overthrown by Marduk (originally
a sun-god), who split Tiamat in half like the shell of an oyster,
making the sky and the sea. Next came the regulation of the solar
system and calendar, the creation of plants and animals, and the
making of humanity.

In Cretan mythology, there is a contemporaneous cosmogony in which Gaia brings forth Earth and Eros from Chaos. The Orphic tradition in Greece is said to derive from Crete. In the oldest Orphic myths, the first principles are Earth, Night, and Heaven. The oldest Greek literature includes Homer (*Iliad, Odyssey*) and Hesiod (*Theogony, Works and Days*). In Homer, there is a cosmogony in which Night is the supreme principle, and Oceanus and Tethys are the father and mother of the gods, including Zeus, modeled on Marduk. It is not certain which of these two traditions is older.

In Hesiod's *Theogony*, written around 700 B.C. (verses 116–136), we find yet another cosmogony, in which Chaos (feminine) is supreme. Here are verses 116–122, in the faithful translation of Athanassakis.

> Chaos was born first and after her came Gaia
> the broad-breasted, the firm seat of all
> the immortals who hold the peaks of snowy Olympos,
> and the misty Tartaros in the depths of broad-pathed earth
> and Eros, the fairest of the deathless gods;
> he unstrings the limbs and subdues both mind
> and sensible thought in the breasts of all gods and all men.[11]

Then came Erebos and Night, Ether and Day, Ouranos and the other gods. This is the first occurrence of the word *Chaos* as far as we know, and its most probable meaning is gap—referring to the gap between the sky and earth, or the gaping void. It did not acquire its current meaning, "any condition or place of total disorder or confusion," at least until the Stoics (500 B.C.). In order to prevent the confusion of Hesiod's Chaos with disorder, it is sometimes translated as Void.[12]

In Hesiod, Gaia means the Earth, and Eros is Desire, the immanent creative energy, the soul of all the unions of the creation story. More abstractly, we may think of Gaia and Eros as Matter and Spirit. For Hesiod, there are three primal cosmic forces: Chaos, Gaia, Eros. From Chaos issue Darkness and Light, Night and Day. From Gaia come Mountains, Sea, and Sky. And from them are born, in turn, all the other deities, in four generations. A later Orphic cosmogony (probably sixth century B.C.) begins with Cronus

(Time), from which Ether and Chaos emerge. It is usually said that the adaptation of the word *cosmos* to indicate the order of the universe originated about this time with Pythagoras. However, it is more likely that this usage originated with Heraclitus or Parmenides, circa 500 B.C.[13]

Finally, we may consider the Hebrew cosmogony from Genesis 1. This chapter is said to date from the preexilian period, before 800 B.C., but it is based on older myths, derived from Babylonia. The Dartmouth Bible enters the first two verses as follows:

> In the beginning God created the heaven and the earth. And the earth was without form, and void; and darkness was upon the face of the deep. And the Spirit of God moved upon the face of the waters.

Here we may recognize several Babylonian influences: (1) "God created the heaven and the earth" recalls Marduk's rearranging of Tiamat; (2) the phrase "without form, and void" (Hebrew *tohu wabohu*) reflects the watery chaos aspect of Apsu and Tiamat; (3) "the waters" (Hebrew *tehom*, which is philologicaly related to *Tiamat*) suggests Mummu, the flood, offspring of Apsu and Tiamat.

Here is a tentative table of equivalences:

Cosmogonical Principles

Babylonian	Hebrew	Greek	English	Connotations
Apsu	Tehom	Oceanus	Deep	father, ocean, god
Tiamat	Tohu	Chaos	Void	mother, ocean, goddess, sea serpent
Mummu	x	x	Flood	son of ocean, mist
Marduk	Yahwey	Zeus	God	law and order, cosmos
x	x	Gaia	Earth	matter
x	x	Eros	Desire	spirit
Anu	x	Ouranos	Sky	heaven
x	x	Tartaros	far west	underworld (Cretan)

In summary, the creation of the universe in our tradition means the subjugation of chaos by cosmos. Ours is a universe of law and order. This tradition dates from Eridu, at least as early as 2000 B.C. Compare the Sanskrit *Rg Veda,* hymn 190 of mandala 10, where there is no contest: "Universal form and harmony were born of cosmic will. . . ." Note that throughout, creation means creation in form, not creation from nothing.[14]

Most cultures have not only mythical creations, but mythical histories as well. The oral traditions of a cultural group seem to coalesce into a consensual reality, or mythical record, filling in the period from creation until the dawn of recorded history. The play of the gods and mythical heros, their family trees and wanderings, are committed to memory by successive generations in the form of epic poems and dramas. In broadest outline, these oral histories are divided into mythical ages of the world.

First, we consider the Babylonian system, probably the source for many narratives of this type. Its cycles and ages were modeled on the astrological zodiac, or path of the sun. The Babylonian zodiacal cycle is divided into twelve signs in four quarters, belonging to Marduk (Jupiter: spring equinox), Ninib (Mars: summer solstice), Nebo (Mercury: fall equinox), and Nergal (Saturn: winter solstice). These represent at once seasons, directions, planets, and gods of the holy trinity (the rulers of the zodiac: Sun, Moon, Venus). The moon (the star of the upper world) and the sun (star of the lower world) are combatants in the Babylonian system. Because of the astronomical phenomenon of the precession of the equinoxes, the zodiacal constellation, or sign, in which Marduk's equinox falls, changes every twenty-two hundred years or so. The period from about 5000 B.C. to 2800 B.C. began in the sign of the Twins, and thus was called the Age of the Twins. Next was the Age of the Bull, then the Age of the Ram. In this progression, virtue decayed from perfection in the first age, the Golden Age, heading for destruction at the end of the 26,400-year cycle of the Great Year (twenty-two hundred years for each of twelve signs). There are many complications due to the frequent calendric reforms made for political reasons by Baby-

Ionian rulers.[15] For the early Greeks, this scheme had apparently decayed into the four declining ages: Golden, Silver, Bronze, and Iron.

One of the most evolved (and best known) of the mythical chronologies is the Hindu doctrine. In the form usually encountered, this has cycles called days of Brahma, or *mahayugas,* each lasting 4,320,000 years (12,000 360-year intervals). Each mahayuga is divided into four ages, or *yugas: krta* (activity), *treta* (third), *dvapara* (second), and *kali* (conflict). This system illustrates two features shared by most of the fully developed chronologies: A creation occurs after each catastrophic annihilation, or end of the cycle, and each age represents a decline in virtue. The length of each yuga is proportional to its virtue. Thus, *krta* has forty-eight hundred intervals, *treta* has thirty-six hundred, *dvapara* has twenty-four hundred, and *kali* has twelve hundred. Length and virtue thus decline in the proportions four, three, two, and one. According to this version, we are now in a *kali* yuga of 432,000 years.[16]

In another version, less well known, the twenty-four thousand-year astronomical cycle is divided equally in two parts, each called a *daiva* yuga (electric cycle). The cycle begins with the precession of the fall equinox into Aries. Then the first half, or declining arc, has four ages, during which mental virtue is lessening: the *satya* yuga (age of truth) of forty-eight hundred years, the *treta* yuga (third age) of thirty-six hundred years, the *dwapara* yuga (second age) of twenty-four hundred years, and the *kali* yuga (age of conflict) of twelve-hundred years. The other half, or ascending arc, likewise has four ages, but they occur in the reverse order, signifying increasing mental virtue. According to this version of the Hindu chronology, we are now in the early phase of an ascending *dwapara* yuga that began around 1700.[17] This version combines periods of declining and ascending virtue into a full cycle, bridging the Asian and the Old Testament traditions.

We can see that our traditional creation myths begin in the middle Neolithic period, around 4000 B.C., the time of the first wave of invasion by the Kurgan people from the northern steppes, the patriarchal domination of the goddess culture, the introduction of the wheel, and the onset of the Periodic Era.[18] The domination of chaos by cosmos, characteristic of *Enuma Elish* and Genesis, co-incides with this important bifurcation.

An important element in *Enuma Elish* is the cosmic battle with a snake-god, a common feature in mythologies throughout the Near East. Conflicts similar to that between Marduk and Tiamat occur in the victory of the storm-god over the dragon in the Hurrian-Hittite story *Illuyanka* and in the daily struggle of the Egyptian sun-god Re with the dragon Apophis.

In the Babylonian myths, the serpent is identified with the disorderly currents in the ocean, which connect with the current meaning of the word *chaos*. As noted earlier, when the word emerged in Hesiod's *Theogony*, it may have meant the gap between the sky and earth, without any suggestion of disorder. But within a few centuries it acquired this meaning, which is definitely part of the original signification of Tiamat. Further, it seems likely that Tiamat developed in Sumer from earlier mythical serpents, representing disorder and creativity in the goddess religion of the early Neolithic.[19] Thus, Hesiod's concept of chaos merged with an earlier tradition, to form our concept of chaos.

When Genesis 1 was written, Chaos and Tiamat reappeared as *tohu wabohu* and *tehom*. But despite the obvious similarities, there are important differences between *Enuma Elish* and *Genesis*. Chief among these is the omission in the biblical narrative of the cosmic battle theme, common to most of the Near Eastern precedents, in which order subdues chaos after a titanic struggle. Monotheism may be the basis for this transcendence of conflict in creation, but remnants of the pagan combat theme do survive in Genesis, where "they practically always appear as a literary device expressing the evil deeds and punishment of the human wicked in terms of the mythical conflict of God with the rebellious forces of primeval chaos."[20]

Here primeval chaos and evil are identified: a bad omen for the essential chaos of life and creativity. Further, the cosmic battle theme may be the source of the mythical concepts of heaven and hell. Apsu and Tiamat, sky and underworld, male and female, order and chaos, heaven and hell, good and evil: all are the same. As

Heraclitus says, "Listening not to me but to the Logos, it is wise to acknowledge that all things are one."[21]

Darwin's classic work *The Creation of the Species* appeared in 1859, while *Enuma Elish* came to the attention of the public in 1875. These two events gave rise to paleontology and the demise of the Genesis cosmogony. Our chronology, based on paleontology, archaeology, astrophysics, historical scholarship, and radiological dating technology, may be compared with the traditional chronologies. The modern version of the ages of the world is firmly established for the recent past but becomes increasingly fictive as we go farther back. The modern theory of the Big Bang, favored by astrophysicists, may be regarded as a creation myth and as a derivative of the Babylonian cosmogony.[22] The whole theory is dependent on evolutionism, a theory that conflicts with the declining virtue aspect of the traditional theories of the ages. Intrinsic to Christianity, evolutionism developed explicitly since Spenser and Darwin, a century ago, along with paleontology. It divides the history of the cosmos into ages, epochs, eras, and eons, in many different ways, depending on the branch of science. In the interesting division proposed by Eric Chaisson, the past has the Era of Energy (a brief flash after the Big Bang), the Era of Matter (the first ten billion years), and the Life Era (about the last five billion years). With luck, an Era of Consciousness will be forthcoming.[23]

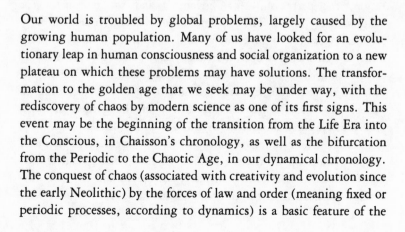

Our world is troubled by global problems, largely caused by the growing human population. Many of us have looked for an evolutionary leap in human consciousness and social organization to a new plateau on which these problems may have solutions. The transformation to the golden age that we seek may be under way, with the rediscovery of chaos by modern science as one of its first signs. This event may be the beginning of the transition from the Life Era into the Conscious, in Chaisson's chronology, as well as the bifurcation from the Periodic to the Chaotic Age, in our dynamical chronology. The conquest of chaos (associated with creativity and evolution since the early Neolithic) by the forces of law and order (meaning fixed or periodic processes, according to dynamics) is a basic feature of the dominator society. To achieve this transformation, we must recognize how, in our day-to-day lives, we help maintain the repression of chaos and creativity.

In our current paradigm, order is to chaos as good is to evil. This has been the status quo for millennia. While culture says disorder is bad, chaos is obviously the favorite state of nature, where it is truly good. But this truth has been banished to the collective unconscious for all these centuries. From the shadows of the unconscious it pushes forth into our consciousness and literature in poetry and song, romance and struggle. It erects heretical monuments in the history of our art, architecture, music, science, and philosophy.

Now, with the aid of chaos theory and large computers, science has discovered the order within chaos. And as chaos becomes acceptable to science, it is seen everywhere, particularly in the life and social sciences, where it invariably provides the substrate of life, evolution, and creativity.

We are learning that chaos is essential to the survival of life. Our challenge now is to restore goodness to chaos and disorder, to replace Tiamat on her rightful throne, in mythology and in daily life, and to reestablish the partnership of cosmos and chaos, so necessary to creation. This will require a major modification to our mythological foundations, unchanged these past millennia: no mean feat.

In *The Chalice and the Blade,* Riane Eisler proposes an anthropological theory in which these are two basic forms of social organization, the partnership (or "gylanic") and the dominator (or "androcratic") forms. The partnership form characterized the early Neolithic period and gave way around 4000 B.C. to the dominator form, which includes both matriarchy and patriarchy. This bifurcation coincides with the discovery of the wheel and the beginning of the Periodic Era now coming to a close. According to Eisler, the peaceful partnership society of the garden of Eden finally disappeared altogether by 1500 B.C. in Minoan Crete. But it lives on in our collective unconscious as a memory. This racial memory wells up from time to time of itself, in "gylanic resurgence waves." For example, the early Christians and the eleventh-century renaissance of the troubadours in the south of France were waves of resurgence of the gylanic culture of Minoan Crete. As Eisler says:

> Moreover, these historical dynamics can be seen from a larger evolutionary perspective. . . . The original cultural direction of our species during the formative years for human civilization was toward what we may call an early partnership, or protogylanic, model of society. Our cultural evolution was initially shaped by this model and reached its early peak in the highly creative culture of Crete. Then came a period of increasing disequilibrium or chaos. Through wave after wave of invasions and through the step-by-step replicative force of sword and pen, androcracy first acted as a "chaotic" attractor and later became the well-seated "static" or "point" attractor for most of Western civilization. But all through recorded history, and particularly during periods of social instability, the gylanic model has continued to act as a much weaker but persistent "periodic" attractor. Like a plant that refuses to be killed no matter how often it is crushed . . . gylany has again and again sought to reestablish its place in the sun.[24]

We now seek to replace dominance with partnership, in a context of psychological and mythological factors deep within the collective unconscious system of global human society.

With chaos and cosmos we have a conflict situation similar to, and related to, the gender-based cultural bifurcation Eisler describes. During the millennia since the beginning of monotheism and the association of chaos with evil in our mythological and religious foundations, there have been revolutionaries of chaos, tossed up into history by "chaotic resurgence waves." Heraclitus (500 B.C.), Jesus (30 A.D.), and Hypatia (350 A.D.) are the best-known chaos revolutionaries of ancient times. More recently, Giordano Bruno (1600), William Whiston (1700), Immanuel Velikovsky (1950), and Wilhelm Reich (1957) stand out. All suffered persecution: crucifixion, burning at the stake, dismemberment, or some such fate.

Naturally, we do not wish to stand out in this way! Perhaps we need not take any intentional action, for we see now that science is in a major upheaval at last, and science is one of the primary watchdogs of the law-and-order domination of society. Its main strategy is to suppress any experience contrary to its dogma, somewhat as organized religion did in the medieval period. Before the Periodic Age, science banished oscillation. Scientists finding oscillation in their laboratory data, in many fields, would jettison the data, as dogma demanded homeostasis. Before the recent dawn of the Chaotic Age, science banished nonperiodic behavior. All disorderly data were called noise and rejected. Now, at last, anything goes. Rest, oscillation, chaotic behavior: All are admitted in the scientific observation of nature.

However, there still exists an evil shadow over chaos. In *Order Out of Chaos,* one of the first books about the reenchantment of science, Prigogine and Stengers say:

> Our vision of nature is undergoing a radical change toward the multiple, the temporal, and the complex. . . . A new unity is emerging: irreversibility is a source of order at all levels. Irreversibility is the force that brings order out of chaos.[25]

They identify reversibility as the disenchanting hypothesis of science and irreversibility as its reenchantment. The role of this reenchantment is creativity and evolution: to bring forth order from chaos. Our brief is different, for we agree with Homer:

> Creation came out of chaos, is surrounded by chaos, and will end in chaos.

That is, order does not come from chaos and leave chaos behind. With no ongoing chaos, there can be no ongoing creation—that is, evolution.

In short, science is rediscovering chaos, and this is seen as a major paradigm shift. Perhaps, with conscious attention, this may evolve into a reenchantment of the world, in which (instead of switching from the domination of chaos by cosmos to the reverse) chaos and cosmos enter into partnership (spiritual gylany along with gender gylany) and we regain the garden of Eden, with our creativity intact: Tiamat rejoined!

NOTES

[1] Abraham and Shaw, *Dynamics, the Geometry of Behavior,* 4 vols. (Santa Cruz, Calif.: Aerial Press, 1982–1988).

[2] Christopher Zeeman, personal communication.

[3] For an account of the role of myth in general evolution theory, see Rupert Sheldrake, *The Presence of the Past: Morphic Resonance and the Habits of Nature* (New York: Times Books, 1988), 255–258.

[4] For a study of this transformation, see Riane Eisler, *The Chalice and the Blade: Our History, Our Future* (New York: Harper & Row, 1987), ch. 4.

[5] Riane Eisler, private communication. See also Merlin Stone, *Ancient Mirrors of Womanhood: Our Goddess and Heroine Heritage,* 2 vols. (New York: New Sibylline Books, 1979) vol. 1, 99–130.

[6] Merlin Stone, vol. 1, 107.

[7] From Eisler, 21, 64. See also Joseph Campbell, *The Mythic Image* (Princeton, N.J.: Princeton University Press, 1974), 77, 157; and Stone, 1979, vol. 1, 82.

[8] From A. H. Sayce, in James Hastings, *Encyclopedia of Religion and Ethics* (New York: Scribner, 1955), 128–129.

[9] Alexander Heidel, *The Babylonian Genesis: The Story of Creation* (Chicago: University of Chicago Press, 1942), 14.

[10] Sayce, in Hastings, 129.

[11] Apostolos N. Athanassakis, *Hesiod, Theogony, Works and Days, Shield: Introduction, Translation, and Notes* (Baltimore, Md.: Johns Hopkins University Press, 1983), 16.

[12] G. S. Kirk and J. E. Raven, *The Presocratic Philosophers* (Cambridge, Mass.: Cambridge University Press, 1957), 26–27. Also, Norman O. Brown, *Hesiod's Theogony* (Indianapolis, Ind.: Bobbs-Merrill, 1953), 56.

[13] Charles H. Hahn, *Anaximander and the Origins of Greek Cosmology* (Philadelphia, Pa.: Centrum, 1985), App. I.

[14] I am grateful to Paul Lee for his many comments on this material.

[15] Alfred Jeremias, in Hastings, 183–187.

[16] H. J. Jacobi, in Hastings, 155–202.

[17] Swami Sri Yukteswar, *The Holy Science* (Los Angeles: Self-Realization Fellowship, 1984), 7–19.

[18] According to Marija Gimbutas, see Eisler, 250.

[19] See Eisler, 86–87. Note that there are different kinds of creation: creation from nothing, creation from within, creation from without. See Eisler, 28, for a similar distinction between actualization power and domination power.

[20] Nahum M. Sarna, *Understanding Genesis* (New York: McGraw-Hill, 1966), 21.

[21] Fragment 118, see Philip Wheelwright, *Heraclitus* (Princeton, N.J.: Princeton University Press, 1959), 102–103.

[22] Rupert Sheldrake, *The Presence of the Past* (New York: Times Books, 1988), 257.

[23] Eric Chaisson, *The Life Era: Cosmic Selection and Conscious Evolution* (New York: Atlantic Monthly Press, 1987), 201.

[24] Eisler, 137.

[25] Ilya Prigogine and Isabelle Stengers, *Order out of Chaos: Man's New Dialogue with Nature* (Boulder, Colo.: Shambhala, 1984), 292.

Big Trouble in Biology: Physiological Autopoiesis versus Mechanistic Neo-Darwinism

LYNN MARGULIS

Central to the autopoietic view is the physiological idea that the material components of all life incessantly move: They cycle at the surface of the earth in chemical transformation and physical transport that always depend directly on the energy from that brilliant star, our sun. Humanity has very little to do with the fact that the matter of life is always transporting and transforming at the surface of the earth. . . . We people . . . accelerate but do not dominate the metabolism of the earth system.

THE CURRENT DILEMMA

More and more, like the monasteries of the Middle Ages, today's universities and professional societies guard their knowledge. Collusively, the university biology curriculum, the textbook publishers, the National Science Foundation review committees, the graduate record examiners, and the various microbiological, evolutionary, and zoological societies map out domains of the known and knowable; they distinguish required from forbidden knowledge, subtly punishing the trespassers with rejection and oblivion; they award the faithful liturgists by granting degrees and dispersing funds and fellowships. Universities and academies, well within the boundaries of given disciplines (biology in my case), determine who is permitted to know and just what it is that he or she may know. Biology, botany, zoology, biochemistry, and microbiology departments within U.S. universities determine access to knowledge about life, dispensing it at high prices in peculiar parcels called credit hours.

As Ludwik Fleck documented, professional knowledge conforms to political realities.[1] Any attempt to breach the acceptable is summarily dealt with, occasionally by devastating criticism but far more frequently by neglect and ignorance. Hence biologists receive Guggenheim Fellowships for calculations of the evolutionary basis of altruism or quantification of parental investment in male children, while the tropical forests are destroyed at the rate of hundreds of acres per day and very little funding exists for study of live plants in their natural environments.

A single example of the current dilemma suffices here: Since the retirement of Professor R. E. Schultes at Harvard University (1986), professional education in the production of food, drug, and fiber compounds by plants from New World tropical forests ("economic botany") is virtually unobtainable in the United States, whereas lessons in Neo-Darwinist religious dogma are exceedingly easy to find. Computer jocks (former physicists, mathematicians, electrical engineers, and so forth), with no experience in field biology, have a large influence on the funds for research and training in "evolutionary biology," so that fashionable computable Neo-Darwinist nonsense perpetuates itself. I here try to explore some of the roots of this institutional malaise.

The big trouble in biology is directly related to big trouble in our social structure and its priorities. This is a big subject. I necessarily limit my comments to the consequences of one philosophical muddle, an aspect of the academic biologists' assumed truth. Science practitioners widely believe and teach—explicitly and by inference—that life is a mechanical system fully describable by physics and chemistry. Biology, in this reductionist view, is a subfield of chemistry and physics. The idea expressed by physicist Sheldon Glashow is commonly held even among biologists: "Just as chemistry is ultimately reducible to physics, so is biology ultimately reducible to chemistry."[2] We compare this pervasive mechanistic belief of biologists, most of whom are smitten by physicomathematics envy, with a life-centered alternative worldview called "autopoiesis," which rejects the concept of a mechanical universe knowable by an objective observer.

Most practicing biologists do not yet know about autopoiesis; as an organized group of scientists, they do not face the issue "What is life?" No tradition in the organization of professional life scientists forces them to ponder life itself. "What is life?" simply is not a subject of inquiry even at plenary sessions of ISSOL (International Society for the Study of the Origins of Life). Rather, biologists, convinced that the universe is mechanical, engage in the incessant search for "mechanisms": of life, the human body, and the environment. By mechanisms they mean sound, light, or chemical signals interacting with carbon-containing matter that determine how life works.

The mechanistic worldview has many problems, one of which is the failure of Neo-Darwinist biologists to think physiologically in general and to recognize the principles of autopoiesis in particular.

Biologists are failing to embrace alternatives to a mechanical universe run by their supposed superiors: physicists, chemists, and mathematicians. A few of the destructive consequences of this philosophy on the academy and its students are outlined here. Both experimental work and theoretical analyses within the life sciences are severely affected by this prevalent physics-centered philosophy. Biochemical research, evolutionary biology, and biological education are all suffering the consequences. Neo-Darwinism is simply one example of a mechanistic philosophy used for illustration in this essay.

AUTOPOIESIS

First, then: What is this alternative to mechanistic Neo-Darwinism? What is this new concept of life, physiological in outlook, called "autopoiesis"? Autopoiesis is a set of some six principles developed by Francisco Varela and colleagues to define the living.[3] *Autopoiesis* combines the Greek words *auto* (self) and *poiesis* (to make); indeed, the latter root also gives rise to *poetry*. It refers to the dynamic, self-producing, and self-maintaining activities of all living beings.[4] The word *autopoiesis* tries to define life by indicating its most indispensable aspects. Properties of autopoietic systems (such as cells, organisms, and communities), along with some physical and chemical correlates of these properties, are listed in the table on page 216. The simplest, smallest known autopoietic entity is a single bacterial cell. The largest is probably Gaia—life and its environment-regulating behavior at the earth's surface.[5] Cells and Gaia display a general property of autopoietic entities: As their surroundings change unpredictably, they maintain their structural integrity and internal organization, at the expense of solar energy, by remaking and interchanging their parts.

"Metabolism" is the name given to this incessant buildup and breakdown of subvisible components—that is, to the chemical activities of living systems. If physiology is the study of the functions of living organisms and their parts, then metabolism is the chemical

Properties of Autopoietic Systems

Property	Aspects	Examples of Biochemical/ Metabolic Correlates
Identity[1]	Structural boundaries; identifiable components; internal organization	Membrane-boundedness; nucleic acid, proteins, fatty acids, and other universal biochemical components of living systems
Integrity[2]/unitary operation[3]	Single, dynamic functioning system	Sum of multienzyme mediated networks and their connection to nucleic acid and protein synthesis
Self-boundedness	Boundary structure produced by system	Lipoprotein membranes; gram-negative, cellulosic, or other cell walls and their connections to primary metabolism
Self-maintenance/ circularity[4]	Boundary structure and components produced by the functioning of the system	Lipogenesis; carbohydrate synthesis; peptidogenesis; nucleic acid synthesis (polymerization); and their interrelations
External supply of component raw materials	External supply of H, C, N, O, S, P, and other elemental constituents	Enzymes that incorporate CO_2, N_2, and so on into cell material: ribulose biophosphocarboxylase (RuBPC'ase), succinyl carboxylase, nitrogenase, and so forth
External supply of energy[5]	Light or chemical energy supply: convertible into organic bond chemical energy.	Chlorophylls: methanogen coenzyme F; bacterial rhodopsin; uptake and incorporation of sugars and other organic compounds into system

[1] Francisco G. Varela, H. R. Maturana, and R. Uribe. "Autopoiesis: The Organization of Living Systems. Its Characterization and a Model." *Biosystems.* 1974, vol. 5, 187–196.

[2] Ibid.

[3] Gail Rainey Fleischaker, *Autopoiesis: System, Logic and Origins of Life* (Boston: Boston University Professors Program, 1988).

[4] Varela et al., op. cit.

[5] Fleischaker, op. cit.

manifestation of those functions. Metabolism can be defined as the sum of the enzyme-mediated network of chemical and energetic transformations of living systems. It is more easily understood as the incessant movements of matter that occur all the time in living systems and that cease when the system dies. Autopoietic systems metabolize, whereas nonautopoietic systems do not. Proteins, viruses, plasmids, and genes are all components of live material. When contained within the boundaries of animal, plant, or other cells, they may be required to sustain cells or organisms and their autopoietic behavior; yet proteins, viruses, plasmids, and genes, intrinsically incapable of metabolism, are never autopoietic in isolation. Metabolism includes gas and liquid exchange (breathing, eating, and excreting, for instance); it is the detectable manifestation of autopoiesis. Autopoiesis determines physiology and hence is the imperative of all live matter. Autopoietic entities—that is, all live beings—*must* metabolize. These material exchanges are the sine qua non of the autopoietic system, whatever its identity. Metabolizing bacteria, of many different types, directly interact with each other via nonautopoietic components (for example, plasmids, viruses). Together, all the bacteria on earth form a worldwide living system—a huge autopoietic entity. Charles Darwin recognized the continuity of the entire system through time, whereas Sorin Sonea has emphasized its unity through space.[6]

Autopoiesis, in principle, does not depend on any specific material substances. Life does not have to be made of water; proteins containing carbon, nitrogen, and hydrogen; nucleic acid—nor any other particular chemical compounds (see table on page 216). However, on earth, since all life today has a common ancestry, all is part of a water–protein–nucleic acid chemical system with continuity for more than 3 billion years. Thus, knowledge of the chemistry of autopoiesis of life on earth provides us a framework to evaluate studies of living beings, especially research on the origin and evolution of life.[7] The autopoietic point of view of dynamic integral systems, using specific carbon-chemical interactions as the basis of self-maintenance, sharply contrasts with the current mechanistic view of life—the parent of Neo-Darwinism, which is so highly fashionable in today's academic circles.

NEO-DARWINISM

Neo-Darwinism, or the "modern synthesis," is a scientific school, primarily in English-speaking countries, that has been in vogue among biologists from universities and colleges since the 1930s. This body of work claims to unite the early twentieth-century discoveries of heredity (transmission, or Mendelian, genetics) with concepts of Darwinian evolution. Mendelian genetics, sometimes disparagingly called beanbag genetics by its critics, is the study of the transmission of traits (eye color, height, enzyme activity) from one generation to another. Evolution, according to the Neo-Darwinist oracle, results from the accumulation of random heritable changes in individuals (mutations).

Gregor Mendel, the monk from Brno, Czechoslovakia, showed definitively that certain heritable traits are indeed transmitted from parents to offspring without dilution, corruption, or any other change. Darwinian evolution, on the other hand, asserts that inherited changes in characteristics of organisms are established in populations as the result of natural selection; it emphasizes the differential survival and reproduction of organisms with distinct hereditary endowment.

Using algebra based on the Mendelian formalism developed for animal populations, Neo-Darwinists proffer formal mathematical explanations for the ways in which organisms evolve. Neo-Darwinism has produced a large body of professional literature that is the sacred text of most evolutionary biologists. Self-identifying Neo-Darwinists control what little funding for evolutionary research exists in this Christian country. Since the seventies, leaning heavily on computer simulations, the Neo-Darwinist religious* movement has generated subfields called population genetics, behavioral ecology, sociobiology, and population biology. The priests and practitioners teach the Mendelian precept that discrete genes act

---

* *Religio*, reverence for the gods, is derived from *re*, again, and *ligare*, to bind people together: hence, *religare*, to bind again.

independently and that the interactions of genes determine the characteristics of the organisms that are selected. Fanciful abstractions have been invented by the Neo-Darwinists, many of whom are scientists who, beginning as engineers, physicists, and mathematicians, found biology "easy." Several of them (for instance, Richard Dawkins of Oxford, Robert Trivers of Santa Cruz, Robert May of Princeton, John Maynard-Smith of Sussex, W. D. Hamilton of Oxford, and George Williams of Long Island) have become famous darlings of life scientists today. I attribute their popularity in part to the soothing effects of their assertions of mathematical certainty.

Yet, as British molecular biologist Gabriel Dover, instructor of genetics at Cambridge University, says, "it is unlikely that true Mendelian genes exist which do not contain any internal repetition and whose mutant alleles rely solely on selection or drift for increased representation in the population."[8] If, as Dover is claiming, the assumptions used by Neo-Darwinists are indefensible, we spectators hardly can expect the mathematics of the subfield biologists listed above to illuminate the histories of life. Those remaining biologists who actually live among and observe metabolizing animals, plants, and microbes have difficulty measuring the quantities or even understanding general concepts labeled and taken as directly observable by the aforementioned mechanistic practitioners (such as "sexual strategy" and "cladistics,"[9] and "inclusive fitness," "evolutionary stable strategies," and "cost-benefit energetics"[10]). These imponderable immeasurables, in my mind, have no reference in the real world. However, the use of such labels serves a crucial social purpose. It binds the users, a growing group of influential scientists and their students, into a cohesive "thought-collective."[11]

Neo-Darwinists, closet Neo-Darwinists, and non–Neo-Darwinists argue among themselves about "who selects" and "what is selected." These intellectual skirmishes become acrimonious.[12] Dover, for example, attempts to extricate us from some of these evolutionary tangles when he writes: "The study of evolution should be removed from teleological computer simulations, thought experiments and wrongheaded juggling of probabilities, and put back into the laboratory and the field. . . . Whilst there is so much more to learn, the neo-darwinist synthesis should not be defended to death by blind watchmakers."[13] (Dover is referring here to the Neo-Darwinist arguments forcefully presented by Dawkins in his 1986 book.[14]) Abner Shimony,[15] in calling natural selection a "null theory," exposes the gross inadequacy of the common oversimplifications. Although the contribution of Darwin himself is lauded and his memory cherished, the physics-centered philosophy of mechanism and its runt offspring Neo-Darwinism[16] is causing the "big trouble" referred to in the title of this essay. Like most scientists, the Neo-Darwinist practitioners see themselves in a simple search for truth, believing they leave philosophy to the philosophers. Of course, they espouse the philosophy in which they are immersed no matter how strongly they protest "neutrality," "objectivity," and "reason."

NEO-DARWINIST OVERSIGHTS

My view is that Neo-Darwinist fundamentals, derivative from the mechanistic life science worldview, are taught as articles of true faith that require pledges of allegiance from graduate students and young faculty members. I include as examples of such fundamentals a nonautopoietic definition of life; a bodiless, linear concept of evolution; and an uncritical acceptance of the mesmerizing concept of adaptation. I paraphrase some of these examples from standard textbooks of genetics and evolution:

> Life, according to the Neo-Darwinist gospel, is a collection of individuals that reproduce, mutate, and reproduce their mutations.

> Evolution, according to this same testament, is change over time in gene frequencies (by gradual accumulation of mutations) caused by natural selection in natural populations.

This standard Neo-Darwinist doctrine asserts that mutations arise by chance. They are chemical changes that are heritable—that is, changes in the DNA sequence of any cell or of any organism comprised of such cells. Such chance mutations, perceived as physical determinants of life that govern the existence of the organism, are purported to be the source of all evolutionary novelty. (Critics of Neo-Darwinism, although they have no well-developed alternatives, have long dismissed the probability that eyes, brains, and flight evolved by chance.[17]) Neo-Darwinists then explain the strong correlation between structures of organisms and their survival requirements with the soothing idea that organisms "adapt" to their environments.

These assertions seem to me to be misdirected, incorrect, or, at best, grossly inadequate. Indeed, the term *adaptation* is used by late twentieth-century biologists exactly as it was by the early nineteenth-century British geologist William Buckland to describe the clever position of the earth in the solar system and the deity's adequacies in his production of durable creations:

> In all these [favorable circumstances] we find such undeniable proofs of a nicely balanced *adaptation* of means to ends, of wise foresight and benevolent intention and infinite power, that he must be blind indeed, who refuses to recognize in them proofs of the most exalted attributes of the Creator. [Emphasis added.][18]

Although philosophers David Abram and Dorion Sagan are among the few to say so explicitly, such prevailing Neo-Darwinist fundamentals with their preevolutionary legacies are frankly at odds with nonmechanistic, including Gaian, system-philosophies of biology.[19] Nonmechanists such as Lovelock, Bermudes, and Dyer incorporate dynamic, interactive physiological thinking whether or not they are explicit about their autopoietic perspective.[20] The life-centered alternatives to mechanistic Neo-Darwinism recognize that, of all the organisms on earth today, only prokaryotes (bacteria) are individuals. All other live beings ("organisms"—such as animals, plants, and fungi) are metabolically complex communities of a multitude of tightly organized beings. That is, what we generally accept as an individual animal, such as a cow, is recognizable as a collection of various numbers and kinds of autopoietic entities that, functioning together, form an emergent entity—the cow. "Individuals" are all diversities of coevolving associates. Said succinctly, all organisms larger than bacteria are intrinsically communities. In this nonmechanistic view, animal and plant physiology becomes a specialized branch of microbial community ecology.[21] Individual animals and plants are not selected by natural selection, since there are no literal "individual" animals or plants; "natural selection" just refers to the fact that biotic potential is not reached; the ability of populations of cells and organisms to maximally grow is always limited by the growth of different cells and organisms and their associated surroundings.

Although appropriately critical biologists such as Dover have reviled the defensive naivete of the "Neo-Darwinist modern synthesis," they have not replaced it with a comfortable philosophical alternative.[22] Hence, insofar as I know, the irreconcilable tensions between the autopoietic and Neo-Darwinist views have not yet been articulated.

FUNDAMENTALISM AND FUNDAMENTALS: THE FLECKIAN THOUGHT-COLLECTIVE

Ludwik Fleck, beginning at the age of forty-seven (in 1943), directed a microbiology and immunology laboratory in Buchenwald until 1945. Saved from the gas showers because he was useful to the Nazis, the Polish Jew Fleck (with his coworkers, primarily Polish physician Marian Ciepielkowski, French serologist and professor X. Waitz, Eugen Kogon, bacteriologist and professor Alfred Balachowsky of the Pasteur Institute, and some German technicians) was put to work producing vaccine against *Rickettsia prowazeckii,* the causative agent of typhus. For two years, while thousands of prisoners were marched to gas chambers just beyond the laboratory doors, Fleck and his colleagues produced large quantities of totally ineffective "vaccine," which was routinely sent to German soldiers at the war zones. Fleck reserved the real vaccine, in exceedingly short supply, to protect himself, his family, and friends. Surrounded by lives in daily danger, Fleck paid close attention to how easily scientists and technicians mentally imbibe the prevalent "common myth." In the end, Fleck's roughly six hundred liters of harmless "vaccine" was never more than a placebo—with which about thirty thousand SS men at the front were injected.

Daily duplicity not only ensured Fleck's survival, but also substantiated his theory of scientific facts. The theory claims that all "scientific facts" are merely consensuses among socially interacting "card-carrying" scientists. Fleck's book develops the concept that "the fact" is a product of a complex social process beginning with individual observation or measurement and terminating with the integration of a stylized "true statement" into the knowledge of the society at large. A practicing microbiologist and scholar for the rest of his life, Fleck—active as a scientist, philosopher, and beloved human being—died in Israel in 1961, some twenty years after his Second World War experiences.[23]

Probably the drama of his own experience confirmed for Fleck the validity of his thesis.[24] A key innovator in the field of the sociology of science, Fleck invented useful methods to analyze scientific activity. He showed how certain words and phrases become banners for the immediate identification of scientific friend or foe. Typical modern-day Fleckian examples include Lamarckism, Lysenkoism, vitalism, mechanism, Darwinism, sociobiology, and even autopoiesis. Fleck documented the processes by which social activities (including attendance at scientific meetings, contributions to professional newsletters and journals, incorporation of common myths into textbooks, and other instruments of socialization) cement into cohesive groups otherwise unruly scientists and technicians. These groups—which Fleck called "thought-collectives"—are then recognizable. They can be evaluated by the process of identification and naming. Once identified and named, the thought-collective achieves the status of "professional tribe," as do today's Neo-Darwinists, whose members are bound together by many ties, including those of common scientific language.

Employing Fleck's concepts, I list in the table on page 224 a small sample of words drawn from Neo-Darwinism in general. Sample Neo-Darwinist terms in current use by molecular evolutionists are listed in the table on the top of page 225. These "technical terms," I claim, have little significance except to the people who identify themselves as members of the scientific disciplines named in the titles of the tables—that is, as members of the thought-collective. By contrast, the universal terms in the table on the bottom of page 225 are concepts relatively independent of language and culture. The value of these quantities are easily measured by scientists now, as they were in the past. Since none of the Neo-Darwinist "battle cries"[25] in the tables below and on the top of page 225 are directly measurable, all quantification associated with them is indirect and necessarily involves various assumptions and unstated hypotheses. These terms, devoid of meaning outside the Neo-Darwinist context, including the molecular evolutionary context, serve this never mentioned quasi-religious purpose: They bind practicing biologists into Fleckian thought-collectives that protect sacred knowledge.

Neo-Darwinism: Words Used as Battle Cries[1]

Adaptation

Altruism, altruistic behavior

Cheating, selfish behavior

Fitness, inclusive fitness

Genetic variation, diversity

Genotype, phenotype

Group selection

Individual

Kin selection

Levels of selection, units of selection, natural selection

Sexual selection, sexual reproduction

Species, race

[1] These kinds of criticisms of Neo-Darwinist concepts and terminology have been made also in David M. Lambert, Craig D. Miller, and Tony J. Hughes, "On the Classic Case of Natural Selection," *Rivista di Biologia—Biology Forum*, 1986, vol. 79, 11–49; and A. J. Hughes and David M. Lambert, "Functionalism, Struetumlism and 'Ways of Seeing,'" *Journal of Theoretical Biology*, 1984, vol. III, 787–800.

Molecular Evolution: Words Used as Battle Cries

Advanced, primitive organisms
Archaeobacteria, eubacteria, metabacteria
Conserved sequences
Eucytes, parkaryotes[1]
Higher, lower organisms
Molecular homology, convergence, divergence
Quickly evolving/slowly evolving molecules
Rooted trees

[1] J. A. Lake, "Origin of the Eukaryotic Nucleus Determined by the Rate-Invariant Analysis of RNA Sequences," *Nature*, vol. 331; pp. 184–186.

Universal Science: Terms and Their Units of Measure

Acceleration (centimeters per second per second)
Density (grams per unit volume)
Energy (ergs)
Heat (calories)
Length (meters)
Light intensity (einsteins)
Magnetism (electromagnetic units per gram)
Mass (grams)
Pressure (torr, atmospheres, bars, millimeters of mercury)
Temperature (degrees Kelvin, degrees Fahrenheit)
Velocity (meters per second)
Volume (length, width, height)
Time (seconds, years)

Why do members of the Neo-Darwinist social group dominate the biological scientific activities in U.S. and other English-speaking academic institutions? Probably there are many reasons, but a Fleckian one is that the Neo-Darwinist mechanistic, nonautopoietic worldview is entirely consistent with the major myths of our dominant civilization. Our rapacious civilization, identified by the fact that international currencies can be exchanged within it, has been characterized by William Irwin Thompson, that master social critic and analyzer of mythmaking (mythopoiesis), as follows:

> We have built up a materialistic civilization that is concerned almost exclusively with technology, power and wealth. . . . Each culture casts its own shadow, a shadow which is the perfect description of its own form and nature.
>
> The shadow which our technological civilization casts is that of Lilith "the maid of desolation" who dances in the ruins of cities. Now that we have made a single polluted city of the entire world, she is preparing to dance in the ruins of our planetary megalopolis. . . .
>
> To effect a reconciliation [with Lilith] man must not seek to rape the feminine and keep it down under him. If he seeks to continue his domination of nature through genetic engineering and the repression of the spiritual, he will ensure that the only release from his delusions can come from destruction. Lilith will then dance on the ruins of Western civilization.[26]

The myths of our technical civilization are easily contrasted with those of some Native Americans. These great people from Beringia (the landmass present some ten thousand years ago, when glaciers bound huge quantities of water in what is now the Bering Strait) preceded all European and African migration onto these two huge American continents. Perhaps we can assume that Chief Seattle speaks for his ancestors and descendants when he says, "The Earth does not belong to Man, Man belongs to the Earth. All things are connected, like the blood which unites us all."[27]

In the world monetary civilization, geological and biological resources are perceived as infinite. Indeed, their very existence is assumed to be determined by human activities (such as market supply, labor, and so forth). Such myths of our technological civilization cannot accommodate an autopoietic-Gaian view of natural history, like that quoted here from Chief Seattle. The Native American perception, just as any nonmechanistic worldview, must be rejected by Neo-Darwinists, in whom such views induce psychic dissonance. A world philosophy based on the recognition of the autopoietic and nonmechanical nature of life *must* upset the believers in the fundamental myths of our technological civilization. In the world of the Native American, humanity belongs to the earth; in the world of the money machines, the earth belongs to humanity. In the autopoietic framework, everything is observed by an embedded observer; in the mechanical world, the observer is objective and stands apart from the observed. In the autopoietic view, the only truly productive organisms are the green photoautotrophs (bacteria, algae, and plants capable of converting sunlight energy into the organic compounds of food) and a few of their bacterial chemoautotrophic relatives (some obscure forms of life, like those living at great depths in submarine vents, capable of converting geochemical energy into food); in the mechanical view, humanity is truly and infinitely capable of being productive. The autopoietic view, which accepts as given that green linen paper is not food and can never be food, also realizes that garbage never goes out, it only goes around; in the mechanical worldview, economics and politics are thought to be directly related to quantities of money and its distribution.

Central to the autopoietic view is the physiological idea that the material components of all life incessantly move: They cycle at the surface of the earth in chemical transformation and physical transport that always depend directly on the energy from that brilliant star, our sun. Humanity has very little to do with the fact that the matter of life is always transporting and transforming at the surface of the earth. The earth behaves physiologically and not mechanically. We people (*Homo sapiens,* only one of perhaps 30 million living species) accelerate but do not dominate the metabolism of the earth system.

We people, for all our architectural maneuverings and hydroelectrical water reroutings, for all our cementing of grasslands and conversion of tropical forests into steak, can never be productive: we can only consume the organic products of the green autotrophs referred to above. Our use of energy for automobile and jet-plane locomotion and our consumption of food such as *Zea mays* (corn) and

Triticum (wheat) is simply the playing out of our autopoietic nature as newly evolved, mammalian-weed apes.[28]

Physiologically oriented biology, studies of life that recognize that autopoietic entities are qualitatively different from other countable matter, tends to be ridiculed or ignored by current practitioners of Neo-Darwinism. I suspect that Neo-Darwinists, upon observing physiology and contemplating autopoiesis, suffer cognitive malaise. Their mathematized formulations systematically ignore physiology, metabolism, and biological diversity; they fail to describe the incessant, responsive, reciprocal effects of life embedded in environment. Suffering philosophical distress, physics-worshiping Neo-Darwinists must reject autopoiesis and its attendant life-centered biology with the same zeal with which the Spanish true church, guarded by its Inquisitors, rejected the mescal- and peyote-eating religions of the Native Americans.

Until the present, only scientists outside the great wall of the English-speaking academy have espoused nonmechanistic, non–Neo-Darwinistic philosophies. Such scientists develop Gaian philosophies[29] or are engaged in building secondary biospheres.[30] In the meantime, inside the monastery, in university life-science departments, victims are accumulating.

Who are the victims of these latter-day religious wars for the souls of the biological science practitioners? Primarily graduate students, young investigators, and teachers, in whom direct observations of life and experience in the field often foster an expansive autopoietic attitude. The study of physiology and immersion, especially in tropical nature, tends to lead students to a perception that the living planetary surface behaves as a whole (the biosphere, the place where life exists on the earth). Yet the academy guards, using Neo-Darwinism as an inquisitory tool, superimpose a gigantic superstructure of mechanism and hierarchy that protects the throbbing biosphere from being directly sensed by these new scientists—people most in need of sensing it. The dispensers of the funds for scientific research and education and other opportunity makers herd the best minds and bodies into sterile laboratories and white-walled university cloisters to be catechized with dogmatic nonsense to such an extent that many doctoral graduates in the biological sciences cannot distinguish a nucleic acid solution from a cell suspension, a sedimentary from an igneous rock, a kelp from a cyanobacterium, or rye from ergot. The English-speaking biology academy has lost sight of the biological priorities. Furthermore, young investigators or students, potential ecologists, botanists, and zoologists who stray from the Neo-Darwinist fold are threatened with expulsion from this prevailing Fleckian thought-collective with its mechanistic thought-style. Were today's budding biologists to take seriously Thompson's mythopoiesis, Varela and Fleischaker's autopoiesis, and Lovelock's Gaian analysis, they would, en masse, have to walk out on the university.[31] In other words, if an individual with ambition to study nature rejects Neo-Darwinist biology in today's ambience, he becomes a threat to his own means of livelihood—that is, to his own autopoietic integrity.

One lesson of the autopoietic concept of biology is that in general, for any organism, many potential threats to its autopoiesis exist. Examples include lack of food, restricted living space, and improper salt balance. A commonly employed name for any general threat to autopoietic integrity is "stress." All organisms—swimming bacteria, surf-battered algae, and hormone-exuding college students—can behave to reduce stress. All organisms respond in ways determined by their hereditary endowment and their environmental astuteness to lessen threats to the self-maintenance of their internal organization. Stress-purging, stress-avoiding, stress-reducing behavior is intrinsic to all autopoietic entities. Nonautopoietic entities do not respond, they are passive. Neither automobiles nor DNA molecules can resist stress.

From these comments it can be concluded that among academic biologists inside the convent walls, Neo-Darwinist reductionism will prevail until the suddenness of a new planetary culture replaces the technological civilization to which Thompson refers. Only after the new civilization binds us consciously to our nonhuman planetmates, especially the truly productive green ones, can the physiology of autopoietic visionaries replace the mechanics of the Neo-Darwinists inside the academic cloister. Alternatively, Neo-Darwinism is expected to prevail until overpopulation (with its concomitant toxic water, polluted airways, and garbage) destroys technological civilization and its money-machine stockpiling thought-collective, of which Neo-Darwinism is only a tiny part.

NEO-DARWINISM AND GAIA

Gaia is the idea that certain environmental surface properties of the earth—for example, temperature and chemical composition of the lower atmosphere—are directly controlled by the biota. (The biota is the sum of the organisms inhabiting the earth: live animals, plants, and microorganisms. The biosphere, which extends some 8 kilometers above and 12 kilometers below the surface of the earth, is the place where the biota resides.) The validity of the Gaia idea, of the self-regulating biosphere, has been forcefully argued by Lovelock.[32] Indeed, the Gaia hypothesis has been called a "grand unified theory" of biology;[33] it also has been recognized as more a point of view than a scientific hypothesis.[34]

In autopoietic language, Gaia is the largest unit we know of that displays the properties listed in the table on page 216. For those unfamiliar with the Gaia hypothesis, probably the best way of thinking about it is to contemplate the assertion that the atmosphere and surface sediments of the earth are part of the living system. That is, life does not *adapt to* a passive physicochemical environment, as the Neo-Darwinists assume. Rather, life actively *produces and modifies* its surroundings. The oxygen we breathe, the humid atmosphere inside of which we live, and the mildly alkaline ocean waters in which the kelp and whales bathe are not determined by a physical universe run by mechanical laws. In stark contrast with a mechanical, physics-centered world, the metabolizing biosphere is physiologically self-controlled. The breathable oxygen, humid air, and mildly alkaline oceans result from the growth of bacteria, plants, and algae that produce oxygen using solar energy; water transportation is driven by the activities of great forests, primarily of neotropical trees; and the neutralization of the acid tendencies of the planet is accomplished by the production of alkaline substances like urea and ammonia by myriad sea creatures (for example, by urination and bad breath).

These are simply three examples of Gaian earth-surface regulatory activities. Many others exist.[35]

The Gaian worldview is an autopoietic one; the surface of this third planet is alive with a connected megametabolism that leads to temperature and chemical modulation systems in which humanity plays a small and only very recent part. (After all, humanity as *Homo sapiens sapiens* evolved only some forty thousand years ago, long after the Gaian system, which is more than 3,000 million years old, was completely in place.)

Neo-Darwinists, who ignore chemical differences between living beings, who never factor autopoiesis into their equations, and who consider organisms as independent entities evolving by accumulation of chance mutations, *must* hate and resist autopoiesis and the Gaian worldview.

If we can assume that consistency is a scientific virtue, then acceptance of a Gaian-autopoietic worldview requires that we reject the philosophical underpinnings of Neo-Darwinism as it is currently practiced. Neo-Darwinism, in the Gaian perspective, must be intellectually dismissed as a minor, twentieth-century sect within the sprawling religious persuasion of Anglo-Saxon biology. As yet another example of a thought-style in the great family of biological-scientific weltanschauungen, past and present, Neo-Darwinism (like phrenology and nineteenth-century German nature philosophy) must take its place (like British social Darwinism) as a quaint, but potentially dangerous, aberration.

The current dilemma, the big trouble of conflicting myths and thought-styles in professional biological science, is not likely to see resolution soon. Speaking for the practitioners of autopoietic-alternative worldviews, who recognize the embeddedness of all people in the great Gaian system, I must applaud the philosophy of Chief Seattle. With him we realize that "Man belongs to the Earth," and money, only green linen paper, is indigestible for all autopoietic entities like us who lack lignases (lignin-digesting enzymes). At the same time, we must face our social fate and scientific destinies. Regrettably, the destinies within academia of the proponents of physiology and autopoiesis probably more resemble those of Seattle and other Native Americans than those that await practicing Neo-Darwinists.

After all, the glorious, greedy tribesmen of western Europe (the aggressors) and their African slaves from whom most people on this green new North American continent are descended and from whom we imbibed our myths of domination, are the true fathers of Neo-Darwinism. These ancestors, sharing a racist and anthropocentric thought-style, easily confiscated the land and decimated the people to replace the nature-knowing Native Americans. Thus, any of us academic biologists who welcome a lively biology should be naive indeed if we conclude that the Neo-Darwinist thought-collective will abrogate its powers and succumb to logic and reason without an intellectual battle to the death. The academic groves and wet field-stations, the university corridors and DNA-recombination laboratories, the governmental funds for mission to planet earth, the ribosomal RNA-sequence data banks, the column chromatographs, the shuttle payload bays, and the contemplation of the Amazon River basin will not be surrendered by the Neo-Darwinists or any other money-machine representatives until a punctuated discontinuity in thought-style penetrates their thought-collective from the outside. Circumstances beyond their control must lead the presently powerful to relinquish their strongholds. Forces beyond their present awareness must overtake these entrenched servants of greedy masters. Perhaps this is what Thompson means when he writes:

> When we have moved beyond the desolation of all our male vanities, from the stock market to the stockpile of rockets, we will be more open and receptive. Open and bleeding like that archaic wound, the vulva, we will be prepared to receive the conception of a new [planetary] civilization. . . .[36]

NOTES

[1] Ludwik Fleck, *Genesis and Development of a Scientific Fact,* English annotated translation (Chicago: University of Chicago Press, 1979).

[2] Sheldon Glashow, *Interactions, A Journey through the Mind of a Particle Physicist and the Matter of this World* (New York: Warner, 1988).

[3] Francisco G. Varela, H. R. Maturana, and R. Uribe, "Autopoiesis: The Organization of Living Systems, Its Characterization and a Model," *Biosystems,* 1974, vol. 5, 187–196.

[4] Gail Rainey Fleischaker, *Autopoiesis: System, Logic and Origins of Life* (Boston: Boston University Professors Program, 1988).

[5] James E. Lovelock, *The Ages of Gaia* (New York: W. W. Norton, 1988). See also notes 30 and 31 below.

[6] Sorin Sonea and Maurice Panisett, *A New Bacteriology* (New York: Jones and Bartlett, 1983): Sorin Sonea, "Bacterial Viruses, Prophages, and Plasmids Reconsidered," *Annals of the New York Academy of Sciences,* 1987, vol. 503, 251–260; and Sorin Sonea, "A Bacterial Way of Life," *Nature,* 1988, vol. 331, 216.

[7] Fleischaker, op. cit.

[8] Gabriel A. Dover, "The New Genesis," in D. L. Hawksworth, ed., *Prospects in Systematics* (Oxford: Oxford University Press, 1988), 151–168.

[9] Colin Patterson, "Cladistics," in John Maynard-Smith, ed., *Evolution Now: A Century After Darwin* (San Francisco: W. H. Freeman, 1983).

[10] John Maynard-Smith, *Evolution of Sex* (Cambridge: Cambridge University Press, 1978); John Maynard-Smith, "Optimization Theory in Evolution," *Annual Review of Ecology and Systematics,* 1978, vol. 9, 31–56; and John Maynard-Smith, ed., *Evolution Now: A Century After Darwin* (San Francisco: W. H. Freeman, 1983).

[11] Fleck, op. cit.

[12] Richard Dawkins, *The Selfish Gene* (Oxford: Oxford University Press, 1976); and Richard Dawkins, *The Extended Phenotype: The Gene as the Unit of Selection* (Oxford and San Francisco: W. H. Freeman, 1982).

[13] Gabriel A. Dover, "Evolving the Improbable," *TREE,* 1988, vol. 3, 84.

[14] Richard Dawkins, *The Blind Watchmaker* (Harlow, England: Longman, 1986).

[15] Abner Shimony, "The Theory of Natural Selection is a Null Theory," in D. Constantini and R. Cook, eds., *Statistics and Sciences* (Norwell, Mass.: Kluwer/Reidel, in press).

[16] Maynard-Smith, *Evolution Now.*

[17] R. W. Clark, *The Survival of Charles Darwin: A Biography of a Man and an Idea* (New York: Random House, 1984); O. A. Reig, "Notes on

Biological Progress, the Changing Concepts of Anagenesis and Macro-evolution," *Genetica Iberica,* 1987, vol. 39, 473–520; N. N. Vorontsov, "The Synthetic Theory of Evolution: Its Sources, Basic Postulates, and Unsolved Problems," *Zhurnal Vses. Khim. Obva im. Mendeleeva,* 1980, vol. 25, no. 3, 295–314.

The origin of radically new behaviors and structures is probably heritable discontinuities that are then modified by mutation (hereditary endosymbioses, karyotypic fissioning, and so forth). These ideas of Neil Todd are detailed in Lynn Margulis, *Symbiosis in Cell Evolution* (San Francisco: W. H. Freeman, 1981), 343–347. L. Margulis and D. Bermudes, "Symbiosis as a Mechanism of Evolution: Status of Cell Symbiosis Theory," *Symbiosis,* 1985, vol. 1, 104–124; and D. Bermudes and L. Margulis, "Symbiont Acquisition as a Neosome: Origin of Species and Higher Taxa," *Symbiosis,* 1987, vol. 4, 185–198.

[18] Quoted in C. C. Gillespie, *Genesis and Geology: A Study in the Relations of Scientific Thought, Natural Theology and Social Opinions in Great Britain, 1790–1850* (Cambridge, Mass.: Harvard University Press, 1969).

[19] D. Abram, "The Perceptual Implications of Gaia: The Gaia Hypothesis Suggests an Alternative View of Perception," *Ecologist,* 1985, vol. 15, 96–103; and Dorion Sagan, "What Narcissus Saw: The Oceanic I/Eye," in John Brockman, ed., *Speculations: The Reality Club* (New York: Prentice Hall, 1990), 247–266.

[20] Lovelock, *The Ages of Gaia;* Margulis and Bermudes, op. cit.; Bermudes and Margulis, op. cit.; B. D. Dyer, "The Nature of the Individual," (in press).

[21] L. Margulis, *Symbiosis in Cell Evolution* (New York: W. H. Freeman, 1982).

[22] Dover, "Evolving the Improbable."

[23] R. S. Cohen and T. Schnelle, *Materials on Ludwik Fleck* (Dordrecht and Boston: D. Reidel Publishers, 1986).

[24] Fleck, op. cit.

[25] Ibid.

[26] W. I. Thompson, *The Time Falling Bodies Take to Light* (New York: St. Martin's, 1981). See also, W. I. Thompson, "Walking Out on the University," in *Passages about Earth: An Exploration of the New Planetary Culture* (New York: Harper & Row, 1981).

[27] Quoted in Joseph Campbell, *The Way of the Animal Powers, Vol. 1:*

Historical Atlas of World Mythology (San Francisco: Harper & Row, 1983), 251.

[28] L. Margulis and Dorion Sagan, *Microcosmos: Four Billion Years of Microbial Evolution* (Boston: Allen & Unwin, 1987), 193–234.

[29] Lovelock, *The Ages of Gaia*. See also Sagan, "What Narcissus Saw," 211–213.

[30] Dorion Sagan, "Biosphere II: Meeting Ground for Ecology and Technology," *Environmentalist*, 1987, vol. 7, 271–281; Dorion Sagan, *Biospheres: Metamorphosis of Planet Earth* (New York: McGraw-Hill, 1990).

[31] Thompson, "Walking Out on the University."

[32] J. E. Lovelock, *Gaia: A New Look at Life on Earth* (Oxford: Oxford University Press, 1979); and Lovelock, *The Ages of Gaia*.

[33] Dorion Sagan, "If the Earth Is Alive: The Gaia Hypothesis May Be Biology's Grand Unified Theory," *Earthwatch*, 1988, vol. 7, 14–15.

[34] Sagan, "What Narcissus Saw."

[35] Lovelock, *The Ages of Gaia;* L. Margulis and D. Sagan, *Microcosmos* (New York: Summit, 1986); and Gregory J. Hinkle, "Marine Salinity: A Gaian Phenomenon?" in Peter Bunyard and Edward Goldsmith, eds., *Gaia, the Thesis, the Mechanism and its Implications. Proceedings of the First Annual Camelford Conference on the Implications of the Gaia Hypothesis* (Cornwall, England: Wadebridge Ecological Centre, 1988), 91–98.

[36] Thompson, *The Time Falling Bodies Take to Light*.

Immuknowledge:
The Immune System as a Learning
Process of Somatic Individuation

FRANCISCO J. VARELA
AND ANTONIO COUTINHO

Immunology is about to emerge from the shadow of its original sin, that of being born from the medicine of infectious diseases, and to cast aside its long-dominant paradigm of vaccination—a heteronomous view par excellence. This happens just when the cognitive sciences are waking up from the dominance of the digital computer as their main metaphor. If we are willing to accept the central importance of autonomous process in both the neural and immune networks, they can teach us how we think with our entire body.

A CHANGE OF METAPHORS

The standard role attributed to immunity is to protect the "self" from the assault of infections. The immune system is supposed to produce defenses against invaders, along with surveillance cells that kill the pathogens and keep the self from foreignness, or non-self. Every immunology text will start by defining immunology as the study of such immune *responses*. A late 1988 *Time* magazine cover story on immunity was illustrated with diagrams of platoons of troops ready for battle.

Immunological discourse revolves around military metaphors just as strongly as cognitive science once revolved around (digital) computer metaphors. Our purpose is to introduce here a substantially different metaphor and conceptual framework for the study of immunity, one that puts the emphasis on the "cognitive" abilities of immune events. Although the term *cognitive* will undoubtedly sound too strong to many people, it seems useful here if for nothing else than as a sharp contrast to the military figure of immunity as defense. For the time being, let us agree to use the word *cognitive* in the same (vague) sense in which we can apply it to biological entities such as brains and animal populations, and not exclusively to mental and linguistic human processes.

The alternative view we are suggesting can best be conveyed by the notion of Gaia as introduced by James Lovelock. He claims that the atmosphere and the earth's crust cannot be explained in their current configuration (gas composition, sea chemistry, mountain shapes, and so on) without reference to their direct partnership with life on earth. We all are used to thinking that the biosphere is constrained by and adapted to its terrestrial environment. But the Gaia hypothesis proposes a reciprocal relationship: this terrestrial environment is itself the result of what the biosphere did to it. As Lovelock puts it: We live in the breath and bones of our ancestors.

As a result, the entire biosphere on earth, Gaia, has an identity as a whole, adaptable, and plastic unity, acquired through time in this dynamic partnership between life and its terrestrial environment.

We are not concerned here with the scientific merits of this idea. Let us transpose the metaphor to immunobiology and suggest that the body is like the earth, a textured environment for diverse and highly interactive populations of individuals. The individuals in this case are the white blood cells, or lymphocytes, which constitute the immune system. The lymphocytes are a diverse collection of species, differentiated by the peculiar molecular markers, or antibodies, they advertise on their membrane surfaces. Like the living species of the biosphere, these lymphocyte populations stimulate or inhibit each other's growth. Like species in an ecosystem, they are enormous generators of diversity: The antibodies and other molecules produced by lymphocytes are by far (a millionfold) the most varied collection of molecules produced in the body, and there are exquisite mechanisms to assure the constant change and diversity of those present at all times.

The lymphocytes' network exists in harmony with their natural ecology, the somatic environment of the body; that environment determines which lymphocyte species exist. But, as in Gaia, the existing lymphocytes alter in a radical way every molecular profile in the body. Thus, as adults, our molecular identity is none other than the partnership between the body and its immune system, shaped throughout life in a unique configuration—a microcosmic version of Gaia.

THE INESCAPABLE COGNITIVE ASPECT OF IMMUNE PHENOMENA

Even to fulfill a defensive role, the immune system must exhibit properties that are typically cognitive. To start with, it must have some capacity for the *recognition* of molecular profiles: the shapes of the intruding agents (or antigens), the "foreignness" capable of endangering the bodily integrity of the subject. Next, the system must have a *learning* ability, to recognize and defend itself against new antigens. Then, it must have a *memory,* in order to retain information about the antigens it has encountered, for future comparison with new antigens.

Recognition, learning, and memory are the properties discussed in the current connectionism, or network, approaches to cognitive mechanisms. Such models are normally linked to the brain as their biological counterpart. We argue that the immune system is a cognitive network not only because of properties it shares with the brain, but, more interestingly, because in both cases similar (or at least comparable) global properties of biological networks give rise to cognitive behavior as emergent properties. This makes the immune system a significant voice in the current investigations of basic cognitive mechanisms.

This discussion would not be possible without new results and recently demonstrated trends in experimental immunobiology that emphasize the network aspect of the immune system. This view entails important shifts in immunological practice and applications. We can illustrate this by the well-established observation of the Promethean character of the immune system: It can respond to antigens it has never seen, including those that are human-made, in which case the immune response is not even explainable as some form of evolutionary adaptation. If one were to think of the immune system merely as a genetically programmed repertoire of unrelated responses, it would be necessary to find a specific response to unpredictable events. The immune system would be the kind of general problem solver that, after years of frustrating trials, artificial intelligence has concluded to be an impossibility. Instead, immunology points to very specific processes through which the system's network operates. This network perspective naturally leads to the notion of an autonomous "cognitive" self at the molecular level as the proper view of immune events.

THE EMERGENCE OF THE IMMUNOLOGICAL DOUBLE BIND

The first immunological theories confronted these cognitive issues by ignoring them. The basic assumption was that all antigens would act as instructions to form antibodies against them. Instructionist theories viewed the immune system as being directed entirely from the outside—a heteronomous process—since antibodies would operate like molecular modeling clay. The idea was strongly motivated by the apparent completeness of the immune response—its Promethean character—as demonstrated in 1912 by Karl Landsteiner. His observations made it literally inconceivable that the system could have all possible molecular profiles already inside it; that would be too wasteful, things must have to be guided from the antigenic side directly.

In these theories "the antibody molecule was considered like universal glue, capable of interacting with any antigenic form, to take its complementary form, to remove the antigen and to keep the memory of the 'learned' configuration."[1] The quotation marks around the word "learned" are significant; they suggest that the "cognitive" nature of the process is inevitable, and one must refer to some sort of learning. At the same time, the process involved is evidently far from cognitive. Otherwise, the paper on which one's signature appears could be said to have learned. Furthermore, the lack of cognitive capacities is more evident, in that a reference to an individual entity *for* which the discrimination and learning happens is completely absent. "These theories contained their own death since such a universal dough cannot discriminate between self and non-self antigens."[2]

This statement touches the key issue. It is by now classical in immunology to talk about the self/non-self discrimination. Such discrimination arises because the immune system acts *inside* a body. This simple fact has deep consequences. Until recently, as we will discuss, immunology had followed the same tendency as other areas of cognitive science in considering any form of cognitive capacity to be a form of information processing: Information is supposed to come in, the system is supposed to produce an appropriate response. Such a process is the core of heteronomous approaches, or those that view the system as outer-directed.[3]

This heteronomous scheme has been faithfully followed by immunologists. An antigen comes in, and the appropriate response is the production of an antibody with the resulting removal of the antigen. But what determines how the antibody is formed? Unlike the nervous system, the immune system has no spatially fixed sensory organs. Antibodies circulate freely inside the organism, and they have as much chance of meeting molecules that belong to the organism's tissues (self) as of meeting antigens (non-self). Briefly stated, *there must be a way to recognize what has to be recognized.*

If the reader thinks things are getting a little too complicated, we agree—they are. But it is important to see clearly that the inevitable need to postulate some form of knowing what is to be known makes a simple heteronomous framework for immunity very unsatisfactory; immunity is not merely an automatic response to something coming from outside. Recent immunology (1950 to 1970) has tried, with little success, to circumvent this difficulty while still keeping the heteronomous viewpoint; a more satisfying solution demands a more radical revision toward understanding the immune system as an autonomous network. But let us go step by step and examine a little more precisely what "recognition" could mean in this context.

To say that an antibody "recognizes" an antigen means that it binds chemically to it and by so doing neutralizes it. Admirable economy when it is a foreign molecule, but not when molecules are essential components of the organism. This simple logic has been the reason that immunologists have excluded *a priori* the possibility that antibodies could attach to "self" molecules without triggering consequences typical of autoimmune diseases. Outside these pathological conditions, the organism normally does not manifest deleterious immune reactions against its own tissues. Rather, it exhibits the phenomenon of "tolerance," described in 1900 by Paul Ehrlich as *horror autotoxicus.*

One important idea established gradually was that the recognition of unknown antigens can be to some important extent determined by the imprecision of the binding mechanism itself—that is, an antibody can bind with varying degrees of affinity to a large spectrum of molecular shapes. Thus, a repertoire of 10^5 kinds of antibodies is sufficient to make the tadpole live, while a human possesses more than 10^9 varieties and is also viable. In other words, there are various ways of being *complete* for the task performed by the immune system. This makes the self/non-self issue vastly more complex.

Double discrimination, double recognition: It is necessary to know what is a non-self antigen before knowing what is an antigen; logically, to recognize non-self entails knowing what self is. The difficulty is that, as we have said, recognition entails destruction. This diabolical predicament can be summarized thus: "The classical theories demand . . . the comparison that differentiates between self and non-self structures, while (imposing) the ignorance of the existence of self or the threat of immunological self-destruction."[4] We would like to call this predicament the (classical) "immunological double bind": One cannot defend without recognizing, one cannot recognize without destroying. Like the United States' policy in Vietnam of "destroying to save," one is faced with two incompatible constraints linked in an inextricable fashion.

THE ESTABLISHMENT OF CURRENT DOCTRINE (CLONAL SELECTION)

The next important step in immunological thinking to confront the knot evoked above, while still holding tightly to a heteronomous view of the immune system, was clonal selection theory. This theory, the result of the contributions of Niels Jerne and MacFarland Burnet during the 1950s, dominated immunology until recently, much as the symbolic-computational view of cognition has dominated cognitive sciences since that time. It took quite a while for immunology to loosen the grip of instructionist theories, and it did so reluctantly.

The first main idea, arising from Jerne's work, was that of an antibody repertoire that remains in the body permanently. Jerne proposed that antibody production *precedes* and, in a certain sense, anticipates the coming of the antigen. At the time, given the dominant instructionist theories, this notion was quite inconceivable. Today we know that there are about 10^{20} antibodies with a high degree of diversity and degeneracy of binding, and the notion of an internal repertoire is no longer in doubt.

It remained to be explained how a collection of antibodies that were initially random relative to a given world of antigens could be shaped by it. It was known that antigenic encounters leave a trace in the system: The antibodies that bind to an antigen increase in substantial numbers; this is the key aspect of an immune response. In fact, it is these sorts of phenomena that eventually will force us to come to grips with cognitive properties. Jerne made the most remarkable suggestion to circumvent this difficulty: He invoked Darwin and natural selection. Even if the antigen does not operate as the blueprint or instruction for antibody formation, it can *select* antibodies that are already present, bind to them sufficiently, and cause them to increase in numbers. It was then left for Burnet to propose a specific mechanism whereby this selective process could be embodied in terms of mere lymphocyte traffic. The basic idea was that every lymphocyte carries (and can produce) only one type of antibody, so each antigen would link up a subclass of lymphocyte families, or *clones*. The contact between antigen and clone leads to the proliferation of the cells of that clone, which then leads to an increased production of antibodies of that particular type, thus neutralizing the incoming antigen.[5] In this fashion the lymphocyte and antibody population evolve under the selective pressures of antigens. The immune system is thus not genetically but antigenically determined; the name "antigenic determinant" is still in use.

Clonal selection theory—that is, the ensemble of ideas detailed above—was a brilliant answer to the thorny question of how the immune system operates in the face of unbounded novelty. The cognitive issues here appear under the garb of an evolutionary play: the transposition of one temporal scale to another, and from the outside environment to an organismic inside. Still, the question of tolerance has not been answered.

The answer of clonal selection theory to this perennial problem is simple. The theory postulates that the initial antibody repertoire is not, in fact, complete: It is missing *precisely* those clones that can recognize self molecules. Simple solution shifting the initial problem to a new one, since this could not be done *a priori* through genetic mechanisms, as we have already discussed. Thus, one could invoke neither genetic process nor selective process in the adult to accomplish this necessary pruning. The only viable solution was to leave this intermediate step to the embryo, where clonal selection theory proposed that anti-self clones had to be deleted. This idea is usually expressed by saying that the organism learns the self/non-self discrimination during ontogeny.

Thus the old cognitive issues reappear through the window after being chased out the door. The selectionist model is not sufficient by itself; one is still forced to introduce a process of learning to delimit a self, albeit a process relegated to embryonic life. By which specific mechanisms would this be accomplished? As Burnet says, anti-self clones can be avoided "by assuming that at this stage of embryonic life the antigenic contact leads to cell death."[6] Curious twist of the previous logic: The mechanism that assures the discrimination between self and antigens becomes precisely the opposite of that which discriminates later between different antigens. In the first case, antigen contact leads to cell removal. In the second case, antigen contact leads to antigen removal. The theoretical move here consists of separating the two poles of the immunological double bind, leaving one side of the discrimination to the adult and the other to the embryo. Moreover, the clones to be eliminated in embryonic life are themselves self components, and thus self-destruction is implicit in a framework developed to avoid it. Clearly, the clumsy fit between these two contradictory processes is a matter of mere logical consistency.

Clonal selection provided a rich source of guidance for experimental work. It led Burnet to postulate the possibility of fooling the immune system by introducing cells into an embryo to enable it to tolerate molecules not normally present. Experimentally, it was indeed established clearly that tolerance was learned. However, it was later to be found that tolerance building is not the exclusive domain of the embryo. The adult also can learn tolerance, and hence the learning cannot be boxed into a particular period of time. This finding poses difficult questions for clonal selection theory.

Furthermore, the notion of a repertoire complete except for self determinants is problematic. The apparently innocent "except for" appears demonic considering what we have said concerning the broad range of molecular profiles to which an antibody binds, and which is at the base of the notion of completeness in the first place. The simple notion of anti-self clonal deletion is simple only with the assumption of antibody specificity, which lies deep in the medical origins of immunology; this assumption is strengthened by the success of vaccination procedures in the few instances when these work by inducing a narrow class of antibodies against a pathogen. But a rule of the form "one antibody, one antigen" is certainly wrong. Removing sufficient clones such that there would be no response to self molecules would amount to depriving the animal of its ability to respond to such a huge number of potential antigens as to compromise its completeness. The protective immunity shield would be perforated. We thus see that the formula of a repertoire complete "except for," conceived to solve the dilemma of the two contradictory forms of recognition necessary for self/non-self discrimination, ends up in another form of contradiction: one demanding a precision in recognition of self that is incompatible with the assumption of completeness.

It often happens that over a few decades of dominance the weaknesses of a theory become sharper and sharper. In the case of clonal selection theory, the combination of the theoretical unsatisfactoriness (not always a concern for immunologists) evoked above and a few key empirical observations opened up in the mid-1970s a new perspective, to which we now turn.

TOWARD AN AUTONOMOUS IMMUNE NETWORK

The dilemmas evoked above remain untouched unless one is willing to give up the original notion of *horror autotoxicus*. It is clear today that there are normal, circulating antibodies that bind to many (all?) self molecules, in both embryos and adults. These antibodies cannot be conceived as being *against* self molecules. In fact, while these same antibody types in large concentration may cause autoimmune diseases, at their normal circulating levels they do not.

But something even more important needs to be reevaluated. One cannot forget that antibodies that circulate and are supposed to carry on the self/non-self discrimination are themselves part of the self. This notion implies the existence of antibodies that bind to other antibodies, or anti-idiotypic antibodies. There is now ample evidence that such antibodies exist and, hence, that the circulating elements of free-floating serum antibodies and the antibodies advertised on cell surfaces are not separate individual elements or clones but are tightly meshed with one another, in what must properly be called a *network* organization. Again, this idea stems from the work of Jerne.[7] Thus, it became necessary to see that the system could operate by its own internal dynamics, in what Jerne called a "eigen-behavior" (self-determined behavior) in a dynamical equilibrium. With these ideas, the notion of a heteronomous immune system was deeply questioned. But it is still necessary to change a few other theoretical assumptions before the full consequences of those questions can be seen.

Imagine an antigen entering into the organism. A part of the antigen, its antigenic determinant, will be recognized by a certain antibody. Let us call this molecular profile E (for epitope). In the old framework, we would say that the anti-E antibody is ready to eliminate the E-carrying antigen. Recognition happens only between the two of them, and the antigen keeps its selective role. In the network perspective, this private dialogue is no longer valid—first, because there is a multiple binding between E and several anti-E antibodies, but more significantly because we have to take into account the antibodies that bind to the epitopes of the anti-E's. These, in turn, will have antibodies that bind to their epitopes, and so on. The result is that we will always run into antibody classes that will at least partially resemble the incoming epitope E. Stated more simply: The antigen will be able to enter into the network to the extent that there is already circulating an antibody with a molecular profile sufficiently similar to it, an "internal image." The antigen ceases to be a "determinant" and *becomes a small perturbation in an ongoing network.* This means that its effects will be varied and dependent on the entire context of the network as it is now known to exist.

We see how the heteronomous view of the system is weakened by merely examining the network logic with which it is constructed. Evidently, when the immunologist injects large amounts of an antigen, the immune response seems like a heteronomous response of the system. But the network view brings into focus how this is a highly contrived laboratory situation. Normally we do not receive large amounts of an antigen. We have a small number of various self molecules that change over life and a certain number of molecules we are exposed to through feeding and breathing. In other words, the system is basically an autonomous unit, open to all sorts of modulation that acts to change its internal levels slightly, but it is certainly not a machine to produce immune responses. Thus, for example, animals that are not exposed to any antigen at all from birth (antigen-free animals) develop an immune system that is quite normal, in blatant contradiction to clonal selection theory, which would have predicted an atrophied immune system.

The next important step, then, is to drop the notion of the immune system as a defensive device, built to address external events, and to conceive it in terms of self-assertion, establishing a *molecular identity* by maintaining circulation levels of molecules through the entire distributed network. It is here that the immune system acquires its full dignity and joins in full with the current research on biological networks. As in all of them, a rich interconnected network generates internal levels through distributed processes. More precisely, a dynamical level of antibody–cell encounters regulates cell numbers and circulating levels of molecular profiles. This idea is strictly parallel to that of the species network giving an ecosystem an identity within its environment. The interesting consequence, of course, is that such an ecology of lymphocytes exists within the body, which changes.

The dance of the immune system and the body is the key to the alternative view proposed here, since it is this dance that allows the body to have a changing and plastic identity throughout its life and its multiple encounters. Now the establishment of the system's identity is a *positive* task and not a reaction against antigens. The task of specifying the identity is seen here as both logically and biologically primary; the ontogenic antigenic history is modulated over that identity.

This concept requires that the immune network—like an ecosystem—have a specific learning mechanism. The mechanism entails the constant changing of the components of the network by recruitment of new lymphocytes from a resting pool (this process reaches up to 20 percent of all lymphocytes per day in the mouse). It is this ongoing replacement that provides the mechanism for learning and memory, instead of the better-known learning algorithms for neural networks. In fact, from the theoretical standpoint, the immune system more closely resembles the flexibility sought by current research in the artificial intelligence strategies known as genetic algorithms or classifier systems.

The reader is surely aware that this presentation of the immune system is sketchy and simplified. It leaves aside enormously important issues, such as the different cell classes that cooperate inside the system (a host of distinctions may be made among lymphocytes: for example, T-helper, T-suppressor, small and large B) and the incredible complexity of molecular mechanisms and their genetic controls (MHS restriction markers, somatic hypermutations, and so forth). But our purpose here is to trace some fundamental conceptual outline. In this sense, it is important to understand what is meant today by a network perspective in immunology. The reality of anti-idiotypic antibodies is unquestioned. What is less clear is their importance and significance. "Immune network" means, to most immunologists, a chain of successive anti-idiotypes. The richness of the network process and its emergent properties, so pervasive elsewhere in the study of complex systems and cognitive science, is, however, not well understood. The number of experimental papers that, properly speaking, study immune network problems, can be counted on the fingers of one hand. The theoretical explorations are just beginning, and depend crucially on the willingness to leave behind the view of immunity as defense even when mediated through idiotype network processes and, instead, to learn to see the immune system as establishing a molecular identity—that is, as an *autonomous,* not heteronomous, system.[8]

Let us now turn to the question of how the autonomous network viewpoint deals with the immunological double bind and the eternal self/non-self discrimination issue. In fact, the answer is quite simple. This approach does what the resolution of any paradox entails:

It jumps outside the domain where it is valid. In the case at hand, this means that *the immune system fundamentally does not (cannot) discriminate between self and non-self.* As we have said above, the ongoing network can be perturbed or modulated only by incoming antigens, then can respond only to whatever resembles what is already there. So any antigen that perturbs the immune network is by *definition* an "antigen on the interior" and therefore will modulate only the ongoing network dynamics. Something that cannot do so is simply nonsensical and may well trigger a "reflexive" immune response— that is, one produced by quasi-automatic processes that are only peripheral to the network itself. The old self/non-self discrimination becomes, at this point, a nonsensical distinction.

Normally, antigens enter the body through food or air and are regulated by the multiple loops impinging on them; thus low levels of both the antigens and the binding antibodies are created. This is precisely what happens with self components. All through development, self molecules interact with the immune components in such a way that their levels are kept within bounds because there is an ongoing immune activity incorporating them. Thus, for example, the level of renin, a normally existing hormone, can be shown to be under the regulation of the multiple antibodies normally present in the immune system. Notice that this ongoing phenomenon need not be (and generally is not) a matter of stability; there is too much variety and replacement of components for this to be so. It is rather a matter of *viability*—that is, a constantly changing trajectory that, nevertheless, never goes beyond certain limits (such as explosive amounts of one antibody type, for example). In this sense the immune system is more like a weather system then it is like the nervous system. It is a matter more of constrained patterns of change than of a few stable nodes acquired through experience. This is what we mean by a *positive* assertion of a molecular identity; what we are in the molecular domain and what our immune system is stand in relation to each other as coevolving processes. Again, we confront squarely what seems to be a reproduction of Gaia inside the body.[9]

The reader who is used to thinking of immunity as defense grows impatient. Surely, he says, you must be joking. For instance, if we have a weakened immunity, as in AIDS, we are immediately ravaged by pathogens. To be sure, the system is *also* able to mount an immune response against infection, as when an antigen enters in too large amounts or enters too quickly. These mechanisms are, interestingly, mostly independent of the network processes just described, and it is almost exclusively these "reflex" immune reactivities that have been the concern of classical immunology.

The point is not to deny that defense is possible, but to see it as a limiting case of something more fundamental: individual molecular identity. In fact, multicellular life is possible without immune defense, as in invertebrates. Defensive responses, the center of attention in medical immunology, are secondary acquisitions, much as defensive/avoidance reactions in neural behavior are necessary later variants of the more fundamental task of motion/relationship in multicellular life. Or in the Gaian metaphor, certainly the stability and plasticity of the eco/biosphere has been remarkably successful to cope with, say, large meteoric impacts. But such events were few and far between, and it seems odd to say that ecosystems evolved because of them. Saying that immunity is fundamentally defensive is as much a distortion as saying that the brain is fundamentally concerned with defense and avoidance. Certainly we do defend and escape attack, but this hardly does justice to what cognition must be about—that is, being alive with flexibility.

REMARKS ON THE FUTURE

The implications of this alternative view are multifarious, but for the sake of brevity, let us emphasize three of special significance in our own current work:

Research questions. This viewpoint suggests new questions about the dynamics of immune events, such as population dynamics, network nonlinearities and emergent properties, connectivity issues, and mechanisms of learning. The number of published papers addressing directly such network questions can (literally) be counted on one hand.[10]

Clinical questions. This viewpoint suggests some alternative ways to address old problems in medicine, such as autoimmunity. In fact, if this

Gaian viewpoint is correct, in principle *every* molecular profile in the body is under immune regulation and could be manipulated.[11]

Machine learning. This viewpoint suggests specific mechanisms through which a complex network is capable of adaptive learning in changing environments. Such algorithms can naturally be lifted and embodied in artificial devices, thus providing artificial intelligence with another biological source of metaphors beyond neural networks.[12]

CODA: JOHNNY'S PARABLE

We have been following the conceptual movement of immunological thinking, from its instructionist inception, through clonal selection, to a modern network perspective. This research logic is inseparable from cognitive issues. It is, in fact, fascinating that, at this point, the immune network takes a place alongside neural networks as a source of both mechanisms and explanations for basic cognitive phenomena such as recognition, learning, memory, and adaptability. If one accepts that connectionism and artificial networks are a valid research alternative in cognitive science, then, for the very same reasons, immune activities are cognitive phenomena. We are fully aware, however, that many would prefer to use the word "cognitive" exclusively for phenomena that involve language and reasoning in humans or machines. We fully acknowledge that this use of the word is a defensible one, but it seems equally defensible to see these "higher" processes in continuity with "simpler" ones, such as those studied by connectionists and exhibited by immune networks. We are interested not in the trivial semantic issue, but in the underlying conceptual issues raised by immune events.

A parable allows us a summary.[13] Johnny needs a suit. He is an instructionist, so he goes to a tailor and has his measurements taken, and a suit is cut accordingly. In a second version of the parable, Johnny is a convinced selectionist. So he goes to a department store and tries on different suits until he finds one that fits him rather well. Clearly, if there is enough diversity in the store, when we see Johnny walking out of the store, we cannot tell if he is an instructionist or a selectionist; but department stores cannot have unbounded varieties, so there is no confusion. The parable has yet a third version of Johnny. This one is a true network thinker. He realizes that choosing a suit is not just a matter of going to the store and picking one, since the choices available depend on a vast social mesh that yields a certain number of different styles and shapes on display at a certain moment of the year in his particular country. He knows that his buying a suit does not affect the social fabric of society as even the faintest perturbation. He knows that whether he buys his suit or not, society will keep on churning out a given variety and regulating style and shape by its own internal logic. This global network activity has constraints, to be sure. For one, there must be some form of clothing for Johnny and everybody else. Second, if armies of Johnnys were to demand a specific type of suit, then the whole social fabric would respond by monotonously accepting that particular taste, in the self-fulfilling prophecy of fashion.

The moral of the parable is that for the first Johnny-the-antigen, society (immunity) is merely heteronomous modeling clay for his desire. For the second Johnny-the-antigen, society (a clonally selected immune system) is present to the extent that it provides an initial diversity for selection and subsequent amplification mechanisms. Such a view is compatible with some form of (weak) network structure for the underlying processes. For the third Johnny-the-antigen, the full autonomy of the society has come forth, making it clear that his choices are not the most interesting event; that even his existence is inseparable from the autonomy of the social fabric, which is where our attention should go; and that Johnny is a secondary side of the story.

Immunology is about to emerge from the shadow of its original sin, that of being born from the medicine of infectious diseases, and to cast aside its long-dominant paradigm of vaccination—a heteronomous view par excellence. This happens just when the cognitive sciences are waking up from the dominance of the digital computer as their main metaphor. If we are willing to accept the central importance of autonomous process in both the neural and immune networks, they can teach us how we think with our entire body.

NOTES

[1] Jacques Urbain, "Idiotypic Networks: A Noisy Background or a Breakthrough in Immunological Thinking?" *Annals de l'Institut Pasteur/Immunologie,* 1986, vol. 137C, 57–64.

[2] Ibid., 58.

[3] Francisco Varela, *Principles of Biological Autonomy* (New York: North-Holland, 1979).

[4] Antonio Coutinho, Luciana Forni, Don Holmberg, Frederick Ivars, and Nelson Vaz, "From an Antigen-Centered, Clonal Perspective on Immune Responses to an Organism-Centered Network Perspective of Autonomous Activity in a Self-Referential Immune System," *Immunological Reviews,* 1984, vol. 79, 151–168.

[5] McFarland Burnet, *The Clonal Selection Theory of Acquired Immunity* (Nashville, Tenn.: Vanderbilt University Press, 1959).

[6] Ibid., 58.

[7] Neils Jerne, "Towards a Network Theory of the Immune System," *Annals de l'Institut Pasteur/Immunologie,* 1974, vol. 125C, 373–389; and N. Jerne, "Idiotypic Networks and Other Preconceived Ideas," *Immunological Reviews,* 1984, vol. 79, 5–24.

[8] Varela, Antonio Coutinho, Bruno Dupire, and Nelson Vaz, "Cognitive Networks: Immune, Neural and Otherwise," in Alan Perelson, ed., *Theoretical Immunology,* vol. 2 (New Jersey: Addison Wesley, 1988); and Francisco Varela, Viceute Sanchez, and Antonio Coutinho, "Viable Strategies Gleaned from Immune Systems Dynamics," in Peter Sanders and Brian Goodwin, eds., *Epigenetic and Evolutionary Order in Complex Systems: A Waddington Memorial Symposium* (Edinburgh: Edinburgh University Press, 1988).

[9] Antonio Coutinho and Francisco Varela, "Immune Networks: A Review of Current Work," *Immunology Today* (in press); and Perelson, op. cit., vols. 1 and 2.

[10] See, for example, Inge Lundqvist, Antonio Coutinho, Francisco Varela, and Don Holmberg, "Evidence for the Functional Interactions among Natural Antibodies," *Proceedings of the National Academy of Sciences* (Wash-

ington, D.C.: National Academy of Sciences, 1988, in press), and Coutinho et al., op. cit.

[11] François Huetz, Frederic Jacquemart, Claudia Peña-Rossi, F. Varela, and A. Coutinho, "Autoimmunity: The Moving Boundaries between Physiology and Pathology," *Journal of Autoimmunity* (1988, in press).

[12] Varela, Sanchez, and Coutinho, op. cit.

[13] Adapted and modified from Massimo Piatelli-Palmerini, "Evolution, Selection, and Cognition," in E. Quagliarello et al., eds., *From Enzyme Adaptation to Natural Philosophy* (Amsterdam: Elsevier, 1988).

The Evolutionary Striptease

DORION SAGAN

To peel off the layers of sexual history is to attempt to pierce the mystery of sex at the center of human reproductive being. In the present archaeology of sexuality, each layer that unfolds brings us closer to the amoral innocence of sex's beginnings, exposing beneath the present the wildness of our animal and preanimal ancestors.

I would love to kiss you
The price of kissing is your life

Now my loving is running toward my life shouting
What a bargain, let's buy it
                              —*Rumi (1207–1273)*

I

Sex has many origins: evolutionary, social-linguistic, and unconscious or metaphysical origins that are not really origins at all, since they stand, in a sense, outside time. To peel off the layers of sexual history is to attempt to pierce the mystery of sex at the center of human reproductive being. In the present archaeology of sexuality, each layer that unfolds brings us closer to the amoral innocence of sex's beginnings, exposing beneath the present the wildness of our animal and preanimal ancestors. And yet the unraveling, the shedding of layer after sexual layer in an attempt to gaze back into the evolutionary past, itself proceeds sexually, in the modest realm of signs and images, if not as a fully exhibitionistic mirror of reality. The evolutionary striptease—a sensual mirage—is therefore seductive or titillating with regard to knowledge. On the one hand, the sex lives of our ancestors, such as "oversexed" ape-boys and ape-girls, "stupid" mesmerizing reptiles, and "cannibalistic" cells can be deduced or hypothesized on the basis of comparative anthropology, primatology, herpetology, genetics, paleobiology, and other sciences. On the other hand, this philosophic representation itself becomes an erotic unveiling, keeping separate the desire to know from the pleasure of knowledge as presence. For times past, as the spatial metaphor for time in the Navaho language reminds us,[1] are more visible than times to come: History is not at our backs but in front of us, spread out, however mistily, for our mind's eye to examine and reevaluate. While the future remains opaque, invisible, we can, with evolutionary hindsight—with retrodiction rather than prediction—*see* where we have been. The translucent past is laid out ahead of us in the mirage of an exotic time dancer as she begins to strip; this mirage is more real than any science fiction guesswork about the future of human sexuality. (Nonetheless, gazing on the evolutionary stripper is like seducing and possessing a person in an erotic fantasy: In both cases gratification is temporary, success illusory.)

And now let the spotlight fall, the curtains part, and the show begin. As we watch, the evolutionary stripper begins shedding layers to reveal our sexual reproductive past, through the gray mist, to the rhythm of silent music. She looks like a fashion model, tall and thin with long thighs and makeup carefully applied to her face. As we watch, her clothes come off and she stands naked, a human female of our species. But the exotic dance begins where most striptease leaves off, and she swirls before us into the past. The slender body dissolves, and a plump Paleolithic woman emerges, wearing clothes made of grass and cosmetics of clay. And then the Paleolithic woman fades into a small estrous ape-woman, with receding forehead and thin hips. The ape's pubic region is swollen and her brown buttocks are striped with rouge and purple. Now she is turning again, shrinking into a still hairier, more unfamiliar primate. But why look? Our interest is in the tantalizing movement of exposing, rather than exposure; so let us turn away from this as yet unrated scene.

▭

Like rummaging through his mate's belongings for evidence of another lover, the seeker of clues to ancestral sex seldom has success. And like the vivid scenes conjured up in the mind of a jealous lover, scenarios of ancestral sex lives—the sex lives of our forebears—also rely upon circumstantial evidence and the Oedipal imagination.

But if we cannot storm in and catch our ancestors evolving *in flagrante delicto,* we can study the bodies and behaviors of live organisms for clues to our animal and microbial past.

II

As the stripper peels off the uppermost layer, we see that the human body itself attests to promiscuity in ape-people.

Since reproduction in many of the species ancestral to humanity was sexual, the shape of the present-day human body conforms to the sexual predilections of our ancestors. Darwin recognized that "sexual selection" was a process as potent as natural selection but was caused by adaptation to members of the same or opposite sex rather than to the environment at large. Human males are adapted to each other and to females by having heavy testicles capable of producing many sperm—far more, for example, than gorillas. Secondary sexual characteristics such as women's breasts (apes have nipples but not true breasts), the absence in women of a distinct period of estrus or heat (as chimpanzees have), and the relatively huge penises of men (again, compared with those of the great apes) have all been explained in terms of the complex relations between the sexes. Over evolutionary time our bodies have changed as dramatically as clothing fashions, partly because of the environment, but also because of the needs and tastes of the opposite gender, as well as competition among each gender for select mates. Ancestral males and females could have lived together in beautiful harmony or violent strife—and no doubt did both. Indeed, in terms of the perpetuation of the sexes, it did not matter much what they did, as long as they came together once in a while, mated, and did the minimal amount necessary to produce offspring capable of doing the same.

Darwin stressed the choice of females in sleeping with "the least distasteful males" and the charm and fighting ability of males in competing for a perpetually scarce supply of females. Yet recently, theoretical biologists have suggested that almost as important as the competition among male bodies for possession of females is the competition among sperm for eggs. Other things being equal, if two or more men copulate with the same woman within a period of about a week, an advantage in begetting offspring will accrue to the one who ejaculates the most sperm. So-called sperm competition arises because active sperm from two or more men may be found at the same time within the same female. This sets up the conditions for a kind of contest inside the female even after she has copulated.

And it is mainly sperm rather than eggs that compete: In humans, for instance, the number of sperm cells released in a single ejaculation is some one hundred seventy-five thousand times more than the number of eggs a woman will produce in her entire lifetime. With this disparity, even marginal female promiscuity promotes conditions for a colossal marathon, a great sperm race in which many can enter but only a few win.

Factors favoring fertilization by the sperm of one male over that of competitors include position during sexual intercourse, number and swimming speed of ejaculated sperm, and proximity of the spermatic means of delivery—the penis—to the egg at time of ejaculation. Copious sperm production (gauged by testicle weight), deep penetration, and an elongated penis all presumably advantage males engaged in sperm competition.

The struggle to fertilize the egg is an example of what Darwin called *intra*sexual selection, competition among members of one sex for access to the other. For Darwin intrasexual selection was primarily male–male competition. In *The Descent of Man and Selection in Relation to Sex* he referred to it as "the power to conquer other males in battle." Where as *inter*sexual selection—"the power to charm the females"—results in the evolution of bright color and ornamentation, *intra*sexual selection can produce a huge body, sharp canines, and other natural weapons. But, given active female sex lives, intrasexual selection can also lead to "peaceful" adaptations, such as a large penis and heavy testicles. Darwin may have been too polite to write—or too Victorian to think—about this fascinating possibility.

Comparing the genitalia of men with those of the great apes, one is led to speculate that sperm competition probably played a greater role in the human past than it does today. Nonetheless, it still occurs. Sperm-competition theorist Robert Smith reports of a German woman who bore twins, one of whom was mulatto—the offspring of an American G.I.—and the other of whom was white—the child of a German businessman. This is clear evidence of sperm competition—though in this case the competition ended in a tie, because the woman produced two eggs, and dizygotic twins were the result. Promiscuity, communal sex, prostitution, infidelity—human sexual behavior ranging from dating to rape—set into motion the contest among multitudes of sperm from different males for a single ovum.

The alternative to entering the sperm competition is to bar other males from the contest altogether—by harassing, bullying, or killing them. (For less brutal souls, marriage and elopement may also be options.) These two main strategies—sperm competition, and sperm-competition *avoidance*—seem to be reflected in the bodies of our closest living primate relatives: the chimpanzee, the orangutan, and the gorilla. Chimpanzees, who can be very promiscuous indeed, produce more sperm per ejaculate and have heavier testicles for their body weight than humans. But the big scary gorillas and orangutans have puny penises, tiny testicles, and unimpressive ejaculate volumes. The average gorilla penis measures only three centimeters (barely over an inch) when erect; the average orangutan hard-on is only four centimeters. But that's all it takes. Think of the film *King Kong,* where the giant airplane-swatting ape climbs the Empire State Building with a beautiful screaming girl in the palm of his hand, and you will understand instantly why the bigger and more ferocious the male, the less natural selection equips him for the sperm competition.

Whereas gorilla and orang males are far larger than their mates, chimp males (and, to a lesser extent, human males) are much closer in size to the opposite sex. The difference in body size ("sexual dimorphism") provides more circumstantial evidence that our ancestors and those of chimps were more promiscuous than the ancestors of the gorillas and orangs. Today gorillas are organized in social hierarchies: Dominant male silverbacks, named for their mature silver coats, typically control a "harem" of females, mitigating against competitive ejaculation contests by shifting their considerable weight around. The expressive orange-haired orangutan, a loner, roams through the forests of Borneo. Couples are often isolated from other orangutans. If orangutan ancestors were similarly isolated, large-penised, heavy-testicled offspring would not have reproduced with greater success than their brethren with smaller genitalia, as the opportunities for female orangs to mate with more than a single male (initiating the sperm competition) would have been very limited.

Men have relatively large penises and heavy testicles—indicating that these may have been valuable survival traits in the past. And here we broach a possible relevance for psychoanalysis in speculative biology: Could it be that, despite the large size of human penises, the human male preoccupation with penis size relates to the importance of sperm competition in the human past? It is difficult to imagine gorillas or orangs worrying about their penile size (if at all, males might worry about the breadth of their chests). Nonetheless, according to Barry McCarthy, author of *Male Sexual Awareness,* two out of three men estimate their penis to be undersized. Such worries also figure in feelings of sexual inadequacy, despite the well-publicized sexological finding that female pleasure depends on external stimulation of the clitoris, which is not even directly stimulated by the penis. McCarthy attributes men's prevalent anxiety about their penis size to several factors. First, boys catch sight of their fathers' penises at an impressionable age and worry they won't catch up. Second, glances at other males in locker rooms are made end-on: the other men's penises seem larger because a man looks at his own penis from above, a perspective that makes it seem smaller because of the shift known by artists as foreshortening. Third, penises seen in a flaccid state do vary dramatically in size; erect, there is far less variation: the average human penis measures five to six inches. McCarthy also cites a generalized male reluctance to discuss personal sexual issues such as penis size: Men are more likely to tell women their sexual worries than they are to tell other men; thus, in the absence of female company, the myth that size is crucial to female pleasure takes longer to be debunked.

But is there a deeper cause for concern here? Might not the prominent role played by penis size in the male psyche derive from the advantages large penises gave protohumans in siring offspring? Certainly there can be no doubt about the large role the penis plays in the unconscious universe unearthed by psychoanalysis—from Sigmund Freud's biologically rooted mental universes of penises and punishment for imaginary castrations, to French psychoanalyst Jacques Lacan's structural linguistic revisions in which we react not to the being or having of the penis per se so much as to the "phallic signifier." If language is itself structured as a play of presence and absence mirroring the early horrors of the child's realization that there exist two kinds of people—those with penises and those with vaginas—then how can the sperm-competition theory prevent the unconscious from *creeping* into its discourse?

It may be therefore that ancestral human beings behaved sexually more like chimpanzees than they did like gorillas or orangs. Chimpanzees in heat, like many other female primates, become pink and swollen around the genitals and anus; they undergo estrus. Females in this state are very sexually active. According to Jane Goodall, Flo, a chimp mother of four, and the pioneering anthropologist's favorite object of study, would become very excited when in heat. During estrus Flo would lift up her pink buttocks flirtatiously and copulate with virtually any male except her own sons. She often enjoyed quite a few males in rapid succession. And while human culture, language, and technology dramatically influence our behavior, polypeptide sequencing—the study of detail in proteins—reveals a closer kinship of humanity to chimpanzees than to any other living species on earth. It may be that both we and chimpanzees derive from a single promiscuous ape ancestor.

What may have happened is that societies featuring marriage outcompeted more promiscuous social units. More cohesive internally, monogamous tribes probably were better equipped to wage violence against their less possessive, more free-loving neighbors. When sperm competition declines, violence and possessiveness become more important. In phenomena ranging from jealous rage to organized antiabortion protests we see males exerting or attempting to exert reproductive control over female bodies; such control effectively replaces the big genitalia of sperm competition as a means of ensuring male reproductive success. Losing estrus, with helpless infants to care for, the smartest ape-women would have saved themselves for provider-fathers more adept at sperm-competition avoid-

ance (that is, violence) than at sperm competition. The importance
of sperm competition therefore would have declined. (Ape-men became humans so recently, however, that the sperm-competition
equipment has had little opportunity to disappear.) In this rough
sketch, I propose that "morality" appears as a cultural phenomenon
legislating the behavior of individuals moving away from sperm
competition and promiscuity. Morality strengthens a *society's* potential for organized violence. As Charles Darwin wrote:

> It must not be forgotten that . . . a high standard of morality gives
> but a slight or no advantage to each individual man and his children
> over other men of the same tribe. . . . [But a tribe whose members]
> were always ready to aid one another, and to sacrifice themselves for
> the common good, would be victorious over most other tribes; and
> this would be natural selection. At all times throughout the world
> tribes have supplanted other tribes; and as morality is one important
> element in their success, the standard of morality [will rise by nat-
> ural selection].[2]

Despite Neo-Darwinian protestations to the contrary, there can be
no doubt that Darwin was right in his implication that not only
individuals but groups of individuals evolve, and that as groups they
are subjected to natural selection. Animal bodies themselves are,
after all, groups of integrated cells, just as societies are groups of
interdependent people. Except for disease growth such as tumors,
cells reproduce "morally"—that is, under a tight regimented discipline of physiological control. Despite the evidence of active sperm
competition in the past, individuals in human society today sexually
restrain themselves because of sociocultural norms stretching from
widespread incest taboos to more parochial restrictions such as the
celibacy vows taken by the Roman Catholic clergy. Not only in
human history, but in the appearance of individuality in evolution
generally, restricting reproduction of individuals may be said to give
an advantage to the reproduction, or spread, of the societies to
which those individuals belong.

In the hundreds of thousands of years preceding history, people

were hunter-gatherers: Evolutionary speaking, our sexual psyches
are probably still responding to life in these ancient times. In this
long period prior to civilization, there were far fewer people on
earth. Male anatomy today attests to more promiscuity in the past;
large penises and testicles presumably arose during times of more
intense sperm competition, perhaps in communal fire-using humans
of the aptly named species *Homo erectus*,[3] or maybe even prior to the
hunting-gathering period. In the form of "Lucy," the fossil record
documents a prehuman female pelvis indicative of an upright walking posture. Perhaps already by four million years ago, as upright
four-foot-tall chimp-faced australopithecines, human ancestors had
evolved large genitalia in the context of sperm competition. And
because primate estrus is so widespread, ancestral women also probably went into heat, swelling up pubically and changing colors,
thereby attracting numerous male suitors. Not only loss of estrus
but the appearance of female breasts represents an enigmatic change
in the prehistory of femininity. For, at first glance, breasts (whose
subcutaneous fat has no direct relationship to milk supply) and the
loss of estrus both would repel males, turn them off: To ape-men
both such traits would be associated with pregnancy and hence a
temporary lack of fertility. The loss of estrus and the development
of breasts would have hidden ovulation—and the oval prize of the
sperm competition. But why would the woman—or her body—
want to hide her fertility? Evolutionarily, the "reasons" ape-women
hid ovulation by concealing estrus could have been similar to the
reasons a married woman does not dress provocatively: She avoids
her husband's jealousy and unwanted male advances. So, too, primordial women who concealed estrus would have benefited by escaping unwanted sexual attention while increasing the chances of
commitment from a single, able man. All humans alive on earth
today may come from females who concealed estrus and ape-men
who, though adapted to sperm competition, gave it up, at least
partially, in order to obtain sexual favors from ape-women. And the
ape-women themselves were probably competing—not for sperm,
which was always in abundant supply, but for help with their crying
infants.

III

Censors having decided that the unrated is not filthy enough to be X-rated, voyeurism resumes, and we catch the stripper in another phase of our outlandish act. Now a primate with a vaguely chimplike face, clever eyes, claws, and a tail, the exotic dancer somersaults acrobatically about the floor, transforming as she rolls into an impish-looking reptile about the size of a dog and with an intelligent, if not wholly affectionate, expression.

Peeling off the second layer, we expose, beneath the sexually selected body, the "reptilian" brain—an ancient part of our anatomy, which we share not only with the apes, but with all mammals and reptiles. The R-complex, as it is called, seems conserved—an instinctive, close-to-the-genes control center still infiltrating our rational consciousness, dragging down the angelic to the level of the human, subverting humans into beasts.

If anatomy is destiny (as Freud said), then this phase of the evolutionary striptease is still more tantalizing than what has gone before: Here, the evolutionary stripper sheds the clothes that are her ape and mammal bodies to reveal a brutal, cold-blooded, and calculating reptilian psyche. The reptile brain appears to be fixated on sex and violence, and to rule the ritual and agonistic behavior—"aggression and submission, territoriality, hierarchies, display, threat, fighting, and vocalizations"—of modern reptiles.[4] For now one may cast much of what goes on at frat parties, rock concerts, and military parades as latter-day manifestations of the reptilian brain. The reptile "within us" appears to exist in a predominantly visual realm that is in some sense timeless—the waking state of living reptiles being perhaps similar to our nocturnal dreams. The continued emphasis on sight in the reptilian brain may come from the lack of a keen sense of hearing and smell in most reptiles. Reptiles process vision more in their retinas than in their brains: Their communications and signification would be instinctive, "dumb"—more like a form of sign language or writing than protracted speech. Yet even metaphors of sign language or body language inadequately describe reptilian thought, for it is in the transition from reptile to mammal brains that the mammalian ability to perceive the passage of time probably occurred. Linearity itself—the "origin" of myth, language, and all evolutionary stories—would then begin at a certain point on the very time line that, paradoxically, it produces.

Neurobiologist Paul D. MacLean has pioneered the scientific description of the human brain as "triune," divided into three sections reflecting our evolution from less brainy ancestors. Freud, starting about 1900, developed his "first topography," in which he distinguished between the unconscious, the preconscious, and the conscious. Later, about 1923, he mapped out a "second topography," which distinguished between *das Es, das Ich,* and *das Uberich,* usually translated as the id, the ego, and the superego, though perhaps more faithfully rendered as the "it," the "I," and the "over-I." Ironically—though he stated that "anatomy is destiny"—Freud stressed that his carefully delineated maps were not descriptions of locality but metaphors for the complex workings of the human mind. The biological thinker Jakob von Uexküll used the word *Umwelt* to refer to the cognitive world, the slice of its environment that each species characteristically internalizes. But MacLean in his triune description shows that the human *Umwelt* is really three in one, the physiologically discrete but mentally overlapping worlds of the triune brain.

The most human and evolutionarily recent part of the brain in this schema is the "neomammalian" cerebral cortex, the external gray matter, which presumably gives us language as well as large heads. Beneath this is the paleomammalian brain, which we share with all mammals from horses to hamsters. The paleomammalian brain seems to govern characteristically "mammalian" emotions, such as melancholy and parental tenderness; MacLean even suggests it is implicated in the enlightenment of aesthetic, scientific, or mathematical discovery—feelings that, like religious awe, convince subjects of the correctness of their thoughts. The old mammalian brain is thought to mediate between the neocortex and the even more "primitive" striatal structures below it.

The striatal complex is "a basic part of the forebrain" consisting of "olfactostriatum, corpus striatum . . . the globus pallidus, and satellite collections of grey matter," according to MacLean.[5] It is called the R-complex because of its striking resemblance to the entire forebrain of reptiles. The stain for cholinesterase, an enzyme that breaks down the nerve transmitter substance acetylcholine, vividly colors and delineates the R-complex not only of reptiles, but of birds and mammals; histochemistry reveals the R-complex to be a uniform brain entity. One histofluorescence technique developed in 1959 causes the R-complex to glow bright green, showing the presence in it of the neurotransmitter dopamine. The R-complex also contains an abundance of serotonin, a neurotransmitter implicated in hallucinogenic LSD experience, and of opiate receptors, which are acted upon by synthetic painkillers such as morphine and heroin. In MacLean's view the R-complex is a basic module of vertebrate brain physiology elaborated by evolution. It appears that in the R-complex we are dealing with an evolutionarily very important part of the human anatomy—perhaps even the physiological site of the unconscious mind.

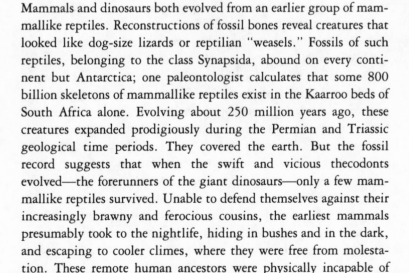

Mammals and dinosaurs both evolved from an earlier group of mammallike reptiles. Reconstructions of fossil bones reveal creatures that looked like dog-size lizards or reptilian "weasels." Fossils of such reptiles, belonging to the class Synapsida, abound on every continent but Antarctica; one paleontologist calculates that some 800 billion skeletons of mammallike reptiles exist in the Kaarroo beds of South Africa alone. Evolving about 250 million years ago, these creatures expanded prodigiously during the Permian and Triassic geological time periods. They covered the earth. But the fossil record suggests that when the swift and vicious thecodonts evolved—the forerunners of the giant dinosaurs—only a few mammallike reptiles survived. Unable to defend themselves against their increasingly brawny and ferocious cousins, the earliest mammals presumably took to the nightlife, hiding in bushes and in the dark, and escaping to cooler climes, where they were free from molestation. These remote human ancestors were physically incapable of competing with the dinosaurs. Those that survived had to evolve expanded sensory modalities—particularly the sense of hearing, which would warn them of the approach of predators and the retreat of prey in their newly nocturnal world. Nonetheless, our sexually reproducing four-legged ancestors must have continued to share with monsters such as *Tyrannosaurus rex* a certain narrow focus on brute survival, on killing, avoiding being eaten alive, and fighting in order to mate, rape, or avoid forced copulation.

Experiments reveal the workings of the R-complex in modern animals ranging from lizards to squirrel monkeys. For example, if one hemisphere of the striatal forebrain of a green *Anolis* lizard is surgically impaired, and one eye is covered while the eye connected to the injured part of the forebrain is left exposed to a rival lizard, the lizard will see his rival but make no ritual display. The connection to his R-complex has been severed; he does not respond with aggressive behavior to sexual competition. But if the other eye is covered—the unimpaired eye connected to the intact part of the R-complex—the reptile reacts with species-typical challenge or territorial displays: he pushes up with his feet, swells out his throat fan, and changes his position so that his imposing long side, his profile, is exposed to his rival. In a word, he makes himself *big*. The normal *Anolis* lizard with intact R-complex is like a man who puffs out his chest or stands over an adversary to get a psychological advantage. He is cold, predictable. You might even call him macho.

Even to us, let alone to the simple reptilian mind, a part can appear as a whole. (If language works by replacing parts for wholes, by synecdoche and similar figures of speech, is it not an elaboration of the displacement and condensation already at work in the reptilian unconscious, in the shifting scene of dreams?) Enlargement of a part, such as puffing out a throat fan or turning one's body to occupy a bigger slice of the rival's field of vision, may be a form of proto-linguistic deception. Such acts trick potential enemies into thinking they see an animal bigger than the one actually present there. Certainly such displays are an economical way to frighten an enemy: You *pretend* to be big.

Of course, it is still more impressive actually to *be* big. In fact, the

whole dinosaur drama, culminating in the extinctions of the dinosaurs some 65 million years ago, is a story of gaining advantage by
increasing size. Could it be that the desire for bigness is coded into
the human R-complex? Translated into English, the R-complex
may contain messages such as "Avoid animals bigger than me" and
"Try to seem as big as I can." Is it possible that the human male
fixation on big penises is a manifestation of a primordial command-
ment or desire for bigness still lurking in the reptilian brain?

But the ancestors of mammals benefited precisely from their
smallness, as it forestalled the effectiveness of the bigger-is-better
strategy. Ultimately, they regrouped and became masters of a different order of bigness: big brains. Appearing big may even have
been important in that other branch stemming from the reptiles—
the birds.

The evolution of birds has long been an evolutionary mystery
because birds could never have evolved wings "in order" to fly; at
first the wings must have been mutant limbs fortuitously valuable
for something other than flight. Some evolutionists suggest that
birds may have used pre-wings and mutant feathery scales as a sort
of insulation, a means of temperature control. But were not wings
also used by birdlike reptiles to frighten their rivals and deceive
them as to their true size? One can imagine that the raising of scaly
wings was an effective premilitaristic display, something like raising
a flag or turning sideways. The wings-to-be could have cast fearful
shadows, petrifying other animals in broad daylight. Suddenly lift-
ing, they could have simulated the approach of much larger beasts
to animals with R-complexes unable to make such distinctions. The
use of such wings would—like a rumbling vocalization by a small
animal to mimic the ground tremors of a larger one or a car suddenly
backfiring on a hot summer day—represent a violence that did not
exist. Flightless, such wings would lie.

Of course primates—thanks to our reptilian endowment—are no
strangers to "ritualistic" behavior. The social behavior of squirrel
monkeys—creatures far closer to us on the evolutionary tree than

ancient reptiles or *Anolis* lizards—was examined in one of MacLean's seminal experiments. Although they do not speak, squirrel monkeys communicate on a preverbal—"physiological"—level. MacLean calls the nonverbal vocal, bodily, and chemical signals in these monkeys "prosematic." He was particularly interested in a gesture that figures heavily in seduction and aggression, in dominance and submission. Male squirrel monkeys use their erections "polysemously" to show aggression, to signal sexual desire, and as a form of platonic greeting. Members of one species invariably hold their erect penises up to their own reflection in a mirror; apparently they are trying to frighten away a rival male. Because they predictably display to their image, MacLean made a systematic study of the effects of brain ablations on their prosematic erection behavior, which he suggests is connected not to more recent parts of the brain, but to the irreducible R-complex. Bilateral lesions of the neomammalian and paleomammalian parts of the primate forebrain caused either no or only a transitory effect on the erect penis display: This kind of behavior was not related to the newer, mammalian parts of the brain. Bilateral lesions of parts of the R-complex, however, "short-circuited" the displays; R-complex–impaired squirrel monkeys no longer exhibited to the mirrors. MacLean was fascinated to find that the squirrel monkeys, apart from not holding their penises up to their images in a mirror, acted virtually normal.

These experiments demonstrate that the intact R-complex is involved in sociosexual behavior not only in reptiles, but in intelligent primates. Outside a certain Edenic or serpentine symbolism, the penis is not usually considered "reptilian." Yet if we look in the evolutionary mirror, monkeys—at least MacLean's squirrel monkeys—are performing a particular ritual of sociosexual exhibitionism at the behest not of their higher mental faculties so much as of their dirty old R-complex. The presence in human beings of an R-complex very much like the generalized forebrain of reptiles from garden lizards to crocodiles suggests that the core of our behavior is still, in a sense, reptilian. However cultivated, orderly, and rational we act, a part of the brain stalks in the shadows like a poisonous, fork-tongued snake.

IV

In the third phase, the evolutionary stripper slips out of her snakeskin skirt and reveals a still more primordial level of sexuality, the ancient cannibalistic gorging, writhing fusions, and miserable duplicity of subvisible cells.

Wet and slippery, sex doesn't fossilize well. Unlike trilobites on Precambrian shores, insects trapped in the precursors of amber, or ape lovers dragging their feet (and thereby leaving footprints) in a romantic stroll through the drying mud, the cellular events at the heart of sex are rarely preserved in the rock record. Indeed, the shedding of clothes by two lovers tends to bring to the surface the warm, wet, salty, sticky environment of the earth as it would have appeared to an observer living 3 billion years ago. Life presumably began in warm, shallow seas; the earliest communities of life were probably sticky mounds, microbial "mats" petrifying into rounded seaside stones called stromatolites. Orgasm psychosomatically returns the body to the softness of its marine origins, to a time when life had not yet hardened, protected, and extended itself by incorporating durable substances such as lignin, shell, and bone.

Some of the oldest unmetamorphosed rocks on earth contain fossils of cells caught in the act of dividing. The South African fossil beds bear rocks that, cut into thin sections and viewed under the high-power light microscope, reveal traces of cell reproduction by fission, or division. But such fission is virtually the opposite of fertilization, or sexual fusion. No fossil yet found preserves the intricate details of mitosis (cell division that yields a perfect copy of the parent cell) or meiosis (cell division yielding sperm or eggs—with only half the parental number of chromosomes—preparatory to fusion). It is because of this second process, meiosis, that we must seek out and join up with the opposite sex to make a new fused, or fertilized, cell if our genes are to be represented in the next generation.

Except for red blood cells (which have nuclei and then extrude them), all the cells in the human body have nuclei, or are eukaryotic. Nucleated cells are characteristic of those that—copied millions of times—make up the bodies of all plants, fungi, and animals. Protoctists—another whole kingdom of life, comparable in scope to plants or animals—are also made of nucleated cells. And the protoctists—slime molds, colonial algae, ciliates, and so on—are still undergoing the sort of cellular experimentation that led to the origins of meiotic sex in the ancestors of plants, fungi, and animals. Evolutionarily, protoctists were clearly ancestral to plants, fungi, and animals, just as bacteria, tiny cells upon which all other life is thought to be based, were ancestral to protoctists.

Since meiotic sexuality is present in some protoctists, absent in others, and represented in a sort of half-hewn or midway state in still others, it seems clear that the sexuality that extends to the human and brings man and woman together began in these unicellular and multicellular organisms, which are usually invisible to the naked eye.

The question then becomes, How did the protoctists—these microbes, more complex than the bacteria but less complex than the colonies that became the first animals—ever hit upon the trick of doubling up their nuclei, chromosomes, and genes every generation? And what is the point of this doubling, when the doubled cell is only to divide in half again? Evolutionary biologists have debated such questions at great length. The potential answers are complex and technical, often revolving around the idea that meiotic sexuality must have some great benefit—for example, that it must somehow "speed up" evolution or have another function that keeps it from disappearing.

In fact, aspects of protoctist sex—meiosis and fertilization—are inseparable from animals as we know them. While all-female species of lizards and rotifers reproduce parthenogenetically—or by the "self" alone—even they remain committed to the prophase part of meiosis. Such virgin-birthing animals in effect fertilize themselves, but they too always make cells that return to the single, so-called haploid, condition each generation. So meiotic sexuality may not have a "reason"; it may simply be an indispensable part of our sexually reproducing being. But, assuming that all organisms originally cloned themselves, how did we get stuck as multicellular, sexually reproducing beings?

Observations by L. R. Cleveland (1892–1969) at Harvard University led him to reconstruct a scenario for the origin of meiotic sex. Meiotic sex may have begun when ancestors of animal cells got caught in cycles of eating each other, doubling, and then dividing again. Theoretically, these protoctist ancestors were cannibals, tempted by starvation and their succulent neighbors. During the eating part of the cycle, a cell would devour a member of its own species, a conspecific, as many do today if starved. But the cannibalism would be partial: As Cleveland actually observed in contemporary protoctists called hairy mastigotes, only the nutritious inessential parts, not the genes and chromosomes of the nucleus, would have been digested. Then the nuclear membranes merged to form a single membrane, and the two-in-one cells remained alive and even were aided in certain environments by their incomplete cannibalism and doubled condition.

In the second part of the cycle, the doubled cell divides meiotically—without reproducing its genes first. The result of this "mistake" in cellular timing would, in some cases, be two "halved" cells, each now with one set of chromosomes, as in the ancestral state (but with a different combination.) Such cycles, though useless in themselves except as temporary adjustment to changing conditions such as seasons and drought, would have been important in the genesis of the cell colonies that were evolving to become the first animals.

Today animals cannot dispense with the return every generation to cells with only one set of chromosomes—the sperm and egg cells that still resemble free-living protoctists, what with their undulating sperm-taillike appendages and having only one set of chromosomes.

V

The evolutionary stripper now discards her glittery top to reveal a deeper, fourth level of sexuality—consisting of liquid patches of bargaining bacteria, promiscuously trading their genes.

Sex in its biological sense exists even beyond the eukaryotic level of a primordial duality when eating and fertilization were perhaps the same sensation. Beyond the polymorphous perversity of the undigested nucleated cell are the symbiotic adventures of bacteria. The nucleated cell itself comes from different types of bacteria symbiotically and quasi-sexually merging into new entities. In nature, bacteria attack, attach to, and penetrate each other; and, living in dense collectives, under the widest variety of conditions, they often continue to trade their genes. For example, genes for photosynthesis have been found outside their milieu, inside parts of cells known as mitochondria, where they serve no conceivable purpose; molecular biology attests that parts of bacteria have roamed, that genetic interchange occurs not only *between* but *within* organisms. Embryology, epigenesis, ontogeny— the whole adventure of individual growth from zygote to sexually mature adult—is a kind of ecological self-organization of "moral" bacteria that takes place in and as the organism. Jumping genes, "redundant" DNA, nucleotide repair, and many other dynamic genetic processes exploit the same ancient bacteria-style sexuality that evolved long before plants or animals appeared on earth. It arose, perhaps, from systems of DNA repair that evolved in cells damaged by solar radiation. Bacterial sexuality fundamentally differs from the sexuality of so-called higher organisms, because it occurs independently of reproduction, crosses "species" barriers, and involves, in principle, the sexual sharing of genes by bacteria all over the world. Indeed, Canadian bacteriologist Sorin Sonea points out that bacteria, since they are able to trade genes freely across would-be species barriers, are not really divisible into (or assignable to) species at all. Instead they form a global "superorganism" whose bodily contours are those of the biosphere itself. If so, this superorganism must be considered "sexual" because it is continually trading genes—although this does not lead to any offspring as we think of them in sexually reproducing species. Sonea presents the global community of bacteria almost as if it were an immortal god: Why would it need children? This masturbating superorganism has already survived for the last four billion years, producing, among other fantasies, humanity.

VI

In the fifth and final phase, the evolutionary stripper takes off everything.

The evolutionary stripper is a curious creature: Her G-string is
not the thin cloth decorated with tassels used by real dancers but
rather a word, a letter, or a musical symbol for ultimate nakedness.
Paradoxically, when she takes off this G-string—to the accompani-
ment of strange vibratory music, consisting in part of a silent triangle and a thunderous crash of cymbals—her nakedness itself is
seen to be gone. She stands before us as fully dressed as ever before.

It might seem that with the genetic exchange in the most ele-
mental living beings, bacteria, we reach the end of the evolutionary
striptease. This, of course, is not the case; not only the nature but
even the possibility of any such final revelation is in doubt. In the
beginning was the Word; and the Word was made Flesh, states the
New Testament. Undoubtedly there is truth to this. For throughout
the entire parade of ancestors—through the whole masquerade of the
evolutionary striptease—we have depended upon a type of information completely different from molecular biology's genes but no less
essential: words. We have encountered not really our sexual ancestors, but only the slippery territory of signs, the shifting of signi-
fiers. And in the use of words, or signs of any kind, there is
necessarily an obscuring, a kind of absence or replacement of the
thing. We postpone reality to discuss it; without this postponement, this instantaneous replacement of our sexual ancestors (or
things in general) by their signs, there could be no possibility of
language, of signification at all. The smattering of ink on the page
erects its own place, which conceals as much as it reveals. What is
so curious is that this postponement, this deferral of the signified by
the signifier is perpetual: It *is* wherever there is language, which is
everywhere. Breathless, we race, we close in, we approach, we an-
ticipate the moment of culmination. But it never comes, or it comes
too soon, or it is already always coming. Such a movement, appearing forever to near its object without ever reaching it, is the very
movement of desire. Thus, in a peculiar way, language becomes a

permanent replacement for the end that orgasm symbolizes. Already
and forever language has become a kind of little death: lipstick on
the mirror of the dressing room of time. The evolutionary dancer,
the exotic chronicler of our striptease, pulls herself apart in order to
show us the truth of the past. But she cannot do it. Instead, she
finally shows herself for what she is: a paper dress, clothes under
clothes, a nudity of pure words.

NOTES

[1] S. M. Lamb, in *Semiotics in Education: A Dialogue,* Spring 1987 (Clare-
mont, Calif.: College Press Inc.), 21.

[2] Charles Darwin, *Sexual Selection and The Descent of Man,* 1803, quoted in
W. Calvin, *Science,* June 24, 1988.

[3] Robert Smith, "Human Sperm Competition," in Robert Smith, ed.,
Sperm Competition and the Evolution of Animal Mating Systems (Orlando:
Academic Press, 1984), 601–653.

[4] Neil Greenberg and Paul D. MacLean, eds., in *Behavior and Neurology of
Lizards, an Interdisciplinary Colloquium* (Rockville, Md.: U.S. Department of Health, Education, and Welfare, Public Health Service, Alco-
hol, Drug Abuse, and Mental Health Administration, National Institute
of Mental Health), DHEW Publication no. (ADM) 77-491, 1978.

[5] Ibid. 1.

Gödel and Einstein
as Companions

═══════════

HAO WANG

*They were great "philosopher-scientists," a very rare breed
indeed, which appears to have become extinct on account
of the intense specialization, the acute competition, the
obsession with quick results, the distrust of reason, the
prevalence of distractions, and the condemnation of
ideals. The values that governed them are to a large
extent considered to be out of date or at least no longer
practicable in their plenitude.*

From about 1942 to 1955 Einstein (1879–1955) and Gödel (1906–1978) frequently walked together while conversing. This was a familiar sight in the neighborhood of the Institute for Advanced Study in Princeton, New Jersey. Their close friendship has been noted occasionally. But it was primarily a private matter, and there is scarcely any record of their discussions, which were almost certainly undertaken entirely for their own enjoyment.

According to Ernst G. Straus, who was with them a good deal in the 1940s: "The one man who was, during the last years, certainly by far Einstein's best friend, and in some ways strangely resembled him most, was Kurt Gödel, the great logician. They were very different in almost every personal way—Einstein gregarious, happy, full of laughter and common sense, and Gödel extremely solemn, very serious, quite solitary, and distrustful of common sense as a means of arriving at the truth."[1]

They were great "philosopher-scientists," a very rare breed indeed, which appears to have become extinct on account of the intense specialization, the acute competition, the obsession with quick results, the distrust of reason, the prevalence of distractions, and the condemnation of ideals. The values that governed them are to a large extent considered to be out of date or at least no longer practicable in their plenitude. Admiration of them takes on the form of nostalgia for a bygone era, or they are regarded as fortunate but strange and mysterious characters. Their lives and work also suggest questions for somewhat idle and vacuous speculations: What would they be doing if they were young today? What types of cultural, social, and historical conditions (including the state of the discipline) are likely to produce minds and achievements like theirs?

There is a natural curiosity about the life and work of people like them. Much has been said about Einstein, and there are signs indicating that a good deal will be said about Gödel as well. Their exceptional devotion to what might be called eternal truth serves to give a magnified view of the value of our theoretical instinct and intellect. Reflections on their primary value may also provide an antidote to all the work done largely for the sake of keeping oneself busy, supply a breath of fresh air, and even point to the availability of more spacious regions in which one could choose to live and work.

However that may be, I have come upon the topic of Einstein and Gödel accidentally and thought about them only according to my own natural inclinations. During the last years of Gödel's life I happened to have a lot of philosophical discussions with him, partly for the shared purpose of making some of his views more widely known. Some of the early discussions have been communicated in my *From Mathematics to Philosophy*,[2] in a form approved by Gödel. Three years after Gödel's death (in 1978), I gathered enough strength to embark on the task of reconstructing the bulk of the discussions from my very fragmentary notes. One thing led to another, and I ended up writing a full-length book, which has since been broken up into two.[3] It was in the process of preparing this book that I became interested in comparing Gödel with Einstein. Moreover, I have come across extensive references to Einstein in Gödel's letters to his mother, so it is possible to learn more about their relations. This essay is an attempt to give a more unified account by digesting selections from the dispersed reports and comments in that book.

Kurt Gödel was born and raised (through gymnasium) in Brünn, the Moravian city where Gregor Mendel, founder of genetics, had done his work. From 1924 to 1939, Gödel studied and made fundamental discoveries in Vienna. In 1933 he began to visit Princeton, where he worked and lived continuously with his wife from spring 1940 until his death. Gödel's doctoral dissertation of 1929 proved that the existing rules of elementary logic (what he calls the logic for finite minds) are complete in the sense that all propositions that are "true in all possible worlds" are derivable. His most famous paper, written in 1931, demonstrated in a surprisingly sharp form that mathematics is "inexhaustible" by any formal system (or computer program). Given any such system and using its resources, one constructs propositions that are intuitively true but not derivable in it. In particular, the proposition (presumed to be true) that it is consistent can be expressed but not proved in it. These results frustrated David Hilbert's intention of finding significant systems that are complete or at least demonstrably consistent (by fairly concrete intuitions). Gödel's work not only suggested a new direction for trying to prove the mind's ability to surpass the machine, but also supplied the main impetus to, and tools for, the development of a pure theory of computers and computability. Supplemented notably by the theory and practice of Alan M. Turing, such a theory has since flourished in various directions to provide a conceptual backbone for the diversified study of computers and their applications.

In the 1930s Gödel shifted his attention to the richer and more chaotic area of set theory, where there was more room for decisive expansions of our vision. From 1943 on, his primary concern was more explicitly philosophy. He found (in the late 1940s) new solutions to Einstein's field equations that permitted "time travel" (time closed upon itself), revealing a striking contrast between our subjective and objective concepts of time. In addition, he made serious studies of Leibniz and Husserl, published several fundamental papers in the philosophy of mathematics, and accumulated a huge mass of unpublished material that will undoubtedly be investigated by scholars for years to come.

Both Einstein and Gödel grew up and did their best work in central Europe, using German as their first language. In the "miraculous" year 1905, when he was about twenty-six, Einstein published articles on (special) relativity, the light-quantum, and Brownian motion. Gödel accomplished his work on the completeness of logic and the inexhaustibility of mathematics before reaching the same age. Einstein went on to develop general relativity, and Gödel moved to set theory to introduce an orderly subuniverse of sets (the "constructible" ones) that yielded the consistency of the continuum hypothesis and has been thus far the single most fruitful step in bringing some order into the chaos of arbitrary sets. (His work on Einstein's equations followed as a digression and a by-product of his study of the philosophical problem of time and change.) During the last few decades of their lives, both of them concentrated on what are commonly thought to be "unfashionable pursuits": Einstein on the unified theory and Gödel on "old-fashioned" philosophy.

The combination of fundamental scientific work, serious concern with philosophy, and independence of spirit reaches in them a height that has rarely been found, and probably is unique, in this century. The supreme level of their intellectual work reminds one of the seventeenth century, sometimes called the century of genius, when important work was published by Cervantes, Bacon, Kepler, Harvey, Galileo, Hobbes, Descartes, Pascal, Huygens, Newton, Locke, Spinoza, and Leibniz.

One indulgence leads to another. If we pair Einstein with Gödel, why not extend the familiar association of Einstein with Newton by analogy? The riddle is then to look for an x such that Einstein is to Gödel as Newton is to x. The obvious candidates are Descartes and Leibniz. Gödel's own hero was Leibniz, and both of them were great logicians. Moreover, Gödel considered Leibniz's monadology close to his own philosophy. At the same time, the clean and conclusive character of his mathematical innovations may be said to be more similar to Descartes' invention of analytic geometry, and his sympathy with Husserl appears to be closer to Descartes' predominant concern with the methods of a new way of thinking and the beginning of a new type of philosophy. Another likely candidate is Pascal, who, like Gödel, often went against the spirit of his time.

In a 1953 letter to his mother, undoubtedly in response to a question from her, Gödel commented on the burden of fame: "I have so far not found my 'fame' burdensome in any way. That begins only when one becomes so famous that one is known to every child in the street, as is the case of Einstein. In that case, crackpots turn up now and then, who desire to expound on their nutty ideas, or who want to complain about the situation of the world. But as you see, the danger is also not so great; after all, Einstein has already managed to reach the venerable age of 74 years."

Gödel's mother, Marianne (1879–1966), was born in the same year as Einstein. When she heard of her son's friendship with Einstein in 1946, she was greatly fascinated and made Einstein one of the major topics in their correspondence for the next decade.[4] She was an exceptionally lively, broadly cultured, and spiritually independent person. An attractive friend and hostess to many, Marianne enjoyed social intercourse; she read widely and loved music, theater, and sports. She was a gymnast (in particular, an ice skater) when young, traveled a good deal in her later years, occupied herself for years on end with the life of Goethe (possessing a series of books on the Goethe Circle in Weimar), and had already attained an appreciation of "modern" literature around 1900 (particularly the works of Schnitzler and P. Altenberg). She was a good wife, even though hers was not a "marriage of love." There were sympathy and affection between her and her husband, a practical man who achieved success in the textile industry and died prematurely when she was forty-nine (and he not quite fifty-five).

Marianne stood on a cordial and friendly, almost comradely, footing with Kurt and his only sibling, Rudolf (Rudi). Rudi was with her much more (in fact, all the time, from 1944 to 1966). But Kurt was the special one, as can be seen from reports by Rudi. "He had an especially good rapport with his mother, who often played for him his favourite melodies (light music) on the piano." "Of my brother as a child my mother recalled many details, which in her view had already then presaged his development into a world famous savant. Among our acquaintances he acquired, for instance, from about the age of 4, the nickname 'Mr. Why,' because he would always question the reason of everything persistently." After the war, Kurt declined to travel to Europe, and Marianne, accompanied by Rudi, visited him several times in Princeton during the last decade of her life. "Especially these trips to the USA were always a festive occasion for her." In 1966 Marianne was very disappointed because she was too weak to travel to Princeton to be with Kurt on his sixtieth birthday.

Gödel's fame has spread more widely since his death in 1978. By coincidence, the surprisingly popular book *Gödel, Escher, Bach* (by Douglas Hofstadter) was published in 1979. A number of conferences have been devoted to his life and work. The first volume of his collected works has appeared, and additional volumes are expected. A Gödel society has been formed in Vienna. Undoubtedly, the growing attention to him and his work is related to the increasingly widespread applications of computers. For instance, one recent symposium to honor him used the topic "Digital Intelligence: From Philosophy to Technology" as its theme.

It is possible that the connection between Gödel's work and computers is closer than that between Einstein's work and the atom bomb, about which Gödel said (in a 1950 letter to his mother): "That just Einstein's discoveries in the first place made the atom bomb possible, is an erroneous comprehension. Of course he also indirectly contributed to it, but the essence of his work lies in an entirely other direction." I believe Gödel would say the same thing about the connection between his own work and computers.

The "entirely other direction" is fundamental theory, which constituted the central purpose of life for both Gödel and Einstein. The combination of this common dedication, their great success with it (in distinct but mutually appreciated ways), and their drive to penetrate deeper into the secrets of nature undoubtedly provided the solid foundation of their friendship and their frequent interactions. Each of them found in the other his intellectual equal who, moreover, shared the same cultural tradition. By happy coincidence, they happened to have been, since about 1933, thrown into the same "club" (namely, the Institute for Advanced Study).

It is hard to find in history comparable examples of intimacy between such outstanding philosopher-scientists. The friendly relations of Newton with Locke, and of Leibniz with Huygens, were not nearly so close. Faraday and Maxwell admired and complemented each other, but they had little personal contact. Among intellectuals of other types we do find a few famous instances: Goethe and Schiller, Hegel and Hölderlin, Marx and Engels.

Gödel was generally reluctant to initiate human contacts and is known to have been comfortable only with a small number of individuals, especially during his Princeton years. There were undoubtedly a number of people who would have liked to interact closely with him, but few had the confidence and the opportunity to find out enough real common interest with him and to maintain continued discussions or other forms of strong personal relations with him. In the case of Einstein, there was, of course, no problem of confidence and plenty of opportunity. Moreover, both of them had thought exceptionally deeply and expressed themselves exceptionally articulately about science and philosophy, on the basis of a wealth of shared knowledge. There is every indication that they greatly enjoyed each other's company and the conversations. Indeed, the interrelation must have been one of the most precious experiences of its kind.

Oskar Morgenstern, who knew Gödel well and was also acquainted with Einstein (probably through Gödel), wrote to the Austrian government toward the end of 1965 to recommend honoring Gödel on his sixtieth birthday. In this letter he said: "Einstein has often told me that in the late years of his life he has continually sought Gödel's company, in order to have discussions with him. Once he said to me that his own work no longer meant much, that he came to the Institute merely 'to have the privilege to walk home with Gödel.' "

From the letters to his mother it is clear that Gödel valued Einstein's company just as much as Einstein valued his. What is involved here is, I think, a fascinating example of human values, possibly helpful in testing ethical theories such as the part having to do with Mill's "principle of preference," which proposes to guide the ranking of pleasures. More than in the quest for definite results or even in the discussions of personal troubles, theirs may appropriately be considered a "purposeless purpose" based on a "disinterested interest." From a common and ordinary perspective, they would be thought to be engaging in a sort of "useless" activity. Yet their genuine enjoyment reveals in a striking manner a type of value many of us can only dimly see and some of us may have experienced sometimes, only to a more limited extent. Could we perhaps call the underlying value that of pure and free inquiry, usually a solitary affair, as an end in itself? The devotion to this value had undoubtedly much to do with their intellectual achievements. But, as Gödel said about the fruitfulness of his philosophical position,[5] it is presumably also true here that a pretended devotion is less effective than a real one.

After Einstein's death Gödel, in response to an inquiry from Carl Seelig, said that he had talked with Einstein particularly about philosophy, physics, politics, and often also on Einstein's unified field theory, although or just because Einstein knew that Gödel stood very skeptically opposed to him. What is presupposed in this statement is, I am sure, a large region of agreement in their taste regarding the value and importance of questions and ideas. They shared a good deal of knowledge (including judgments on what is known and what is not), as well as a great talent for expressing their thoughts clearly. I would like to contrast their outlooks by stating some of their agreements and disagreements.[6]

Both Einstein and Gödel were concerned primarily, and almost exclusively so in their later years, with what is fundamental. For example, Einstein often explained his choice of physics over mathematics partly in terms of his feeling that mathematics was split up into numerous specialties, while in physics he could see what the important problems were.[7] But once he said to Straus, "Now that I've met Gödel, I know that the same thing does exist in mathematics." In other words, Einstein was interested in problems fundamental to the whole of mathematics or the whole of physics but could initially discern only those of physics. Gödel once told me, almost apologetically (probably to explain why he had so little of what he considered success in his later decades), that he was always after what is fundamental.

Neither Einstein nor Gödel, contrary to the prevalent opinion of the physics community, considered quantum theory to be part of the ultimate furniture of physics. Einstein seems to have looked for a complete theory in which quantum theory is seen as a derivative ensemble description. In physics, according to Gödel, the present "two-level" theory (with its "quantization" of a "classical system," and its divergent series) is admittedly very unsatisfactory.[8]

In the letters to his mother Gödel often explained Einstein's attitude with sympathy. For example, in 1950 he commented on an article calling Einstein's theory the "key to the universe" and said that such sensational reports were "very much against Einstein's own will." He added: "The present position of his work does not (in my opinion) justify such reports at all, even if results obtained in the future on the basis of his ideas might perhaps conceivably justify them. But so far everything is unfinished and uncertain." This opinion, I think, essentially agrees with Einstein's own.

These and other examples of agreement reveal a shared perspective that went contrary to common practice and the "spirit of the time," and constituted a solid foundation for their mutual appreciation. Against this background one might say that their disagreements and differences were secondary. Indeed, in other aspects as well, the opposition of their views can usually be seen as branchings out from a common attitude.

For example, both of them valued philosophy, but they disagreed on its nature and function. They were both very much peace loving and cosmopolitan in outlook, but unlike Einstein, Gödel took no public actions. They were both sympathetic with the ideal of socialism, but Gödel's skepticism toward the prevalent proposals on the ways to attain it contrasted with Einstein's less qualified view expressed in the 1949 essay "Why Socialism?" In some sense, both of them can be said to have been religious, but Einstein spoke of accepting Spinoza's pantheism, while Gödel called himself a theist, following Leibniz. (In 1951 Gödel said of Einstein, "He is undoubtedly in some sense religious, but certainly not in the sense of the church.")

They both read Kant in school and had a strong taste for philosophy when young. Einstein was turned against it by the reigning vagueness and arbitrariness, but Gödel went on to devote a great deal of energy to its pursuit, aiming at "philosophy as a rigorous science." According to Einstein: "Epistemology without contact with science becomes an empty scheme. Science without epistemology is—insofar as it is thinkable at all—primitive and muddled."[9] In contrast, Gödel showed less interest in epistemology and believed that the correct way to do philosophy is to know oneself. For Gödel science only uses concepts, while philosophy analyzes our primitive concepts on the basis of our everyday experience.

In the 1950s Einstein, like most intellectuals at that time, preferred Stevenson over Eisenhower, but Gödel was strongly in favor of Eisenhower. (On the other hand, Gödel was more like his colleagues in being a great admirer of Franklin D. Roosevelt.) Ein-

stein's love of classical music is well known; Gödel, however, had
little interest in it. (In 1955 he attended a memorial concert for
Einstein and said afterward, "It was the first time I let Bach, Haydn,
etc. befall me for two hours long.") On the other hand, Gödel's
reported liking for modern abstract art was presumably not shared
by Einstein. Einstein married twice, had two sons and two step-
daughters, and was a widower for almost two decades. Gödel married only once and relatively late, had no children, and was survived
by his wife.

In the letters to his mother Gödel often reported, "I see Einstein
almost daily." He frequently commented on Einstein's health, usually in optimistic terms. There were explanations of Einstein's public activities and observations on books and articles about Einstein.
There was in 1949 even an exchange of gifts on the occasions of
Einstein's seventieth birthday and the Gödels' housewarming. In
summer 1947 Gödel reported to his mother that Einstein was taking
a rest cure: "So I am now quite lonesome and speak scarcely with
anybody in private." In January 1955 Gödel wrote: "I am . . . not
at all so lonely as you think. I often visit Einstein and get also visits
from Morgenstern and others."

A week after Einstein's death, Gödel wrote that the death of
Einstein had of course been a great shock to him, since he had not
expected it at all, and that naturally his state of health had turned
worse again during the last week, especially in regard to sleep and
appetite. Two months later he said, "My health now is good. I have
definitely regained my strength during the last two months."

In terms of the contrast between participation in (rather than
distancing from) history and understanding the world, both Einstein and Gödel were engaged primarily in the task of understanding; in the process, they contributed decisively to their special
subjects. But unlike Gödel, Einstein also participated in history in
other ways, and he was much more a public figure. Gödel kept a
greater distance from the spirit of the time. He speculated and
offered novel ideas on a number of perennial issues that did not
interest the specialists and were, moreover, shunned by most of
them. For example: Is mind more than a machine? How exhaustive
and conclusive is our knowledge in mathematics? How real are time

and change? Is Darwinism adequate to give an account of the origins of life and mind? Is there a separate physical organ for the handling of abstract impressions? How precise can physics become? Is there a next world?[10] Einstein, I believe, paid much less attention to these questions.

While Einstein concentrated on physics throughout his life, Gödel at first shifted his interest from theoretical physics to mathematics and then to logic and, after his great success in logic, involved himself deeply in several philosophical projects. Even though Einstein left his unified field theory unfinished, Gödel was more likely to embark on new voyages, apparently pursuing several important lines of work without bringing them to completion. One might wish to say that Gödel did not plan his life as well as Einstein did his, and that Einstein had a sounder sense of what was feasible. But then, none of us is equipped to foretell with any assurance what fruits our unfinished work will bear in the future. Moreover, as Gödel says, even though the present time is not a good one for philosophy, this situation may change. More generally, we have a tendency to expect the dominant trend to continue in the same direction, but in fact history is full of swings of the pendulum. We have no solid evidence to exclude the possibility of witnessing the appearance of many other powerfully effective intellects of the type represented by Einstein and Gödel, perhaps even in the not too distant future.

NOTES

[1] G. Holton and Y. Elkena, eds., *Albert Einstein: Historical and Cultural Aspects* (Princeton: Princeton University Press, 1982), 422.

[2] Hao Wang, *From Mathematics to Philosophy* (London: Routledge & Kegan Paul, 1974).

[3] *Reflections on Kurt Gödel* (Cambridge, Mass.: MIT Press, 1987). Parts of the original manuscript have been taken out for an expanded reorganization into a separate book, *Conversations with Gödel,* which is under preparation.

[4] I am grateful to the City Library of Vienna, where the letters from Gödel are kept, for permission to quote from them. The following digression on Marianne is drawn largely from a biography of her, prepared by Gödel's elder brother, Rudolf (1902–), in 1967 and expanded in 1978, after Gödel's death.

[5] Wang, *From Mathematics to Philosophy,* 8–11.

[6] The remainder of this essay summarizes my detailed report in *Reflections on Kurt Gödel.*

[7] Compare P. A. Schilpp, ed., *Albert Einstein, Philosopher-Scientist* (Chicago: Open Court, 1949), 15, and H. Woolf, ed., *Some Strangeness of Proportion* (Princeton: Princeton University Press, 1980), 485.

[8] Wang, *From Mathematics to Philosophy,* 13.

[9] See Schilpp, op. cit., 684.

[10] Gödel's views on some of these questions are reported in *From Mathematics to Philosophy,* 324, 326, and 385. More extensive reports of his views on all these questions are given in the two books mentioned in note 3.

Contributors

Ralph H. Abraham is a professor of mathematics and director of the applied and computational mathematics program at the University of California at Santa Cruz.

Antonio Coutinho is a biologist at the Unite de Immunobiologie, Institut Pasteur, in Paris.

Paul Davies, professor of theoretical physics at the University of Newcastle upon Tyne, is author of *God and the New Physics* (Simon & Schuster), *Superforce* (Simon & Schuster), and *The Cosmic Blueprint* (Simon & Schuster).

K. Eric Drexler is a researcher at MIT and Stanford who is concerned with emerging technologies and their consequences for the future. He is the author of *Engines of Creation* (Anchor).

Gerald Feinberg, a particle physicist, is professor of physics, Columbia University, and author of *Life Beyond Earth* (with Robert Shapiro; William Morrow) and *Solid Clues* (Simon & Schuster).

Kevin Kelly is publisher and editor of *Whole Earth Review.*

James E. Lovelock, independent scientist, is president of the Marine Biology Association, a fellow of the Royal Society, London, and author of *Gaia: A New Look at Life on Earth* (Norton).

Lynn Margulis is a biologist, Distinguished Professor (Department of Botany) at the University of Massachusetts at Amherst, and author of *Symbiosis in Cell Evolution* (W. H. Freeman), *Origins of Sex* (with Dorion Sagan; Yale), and *Microcosmos* (with Dorion Sagan; Summit).

Richard Morris, theoretical physicist, is author of *Dismantling the Universe* (Simon & Schuster), *Time's Arrows* (Simon & Schuster), *The Nature of Reality* (McGraw-Hill), and *The Edges of Science* (Prentice Hall Press).

Dorion Sagan is a writer, author of *Origins of Sex* (with Lynn Margulis; Yale), *Microcosmos* (with Lynn Margulis; Summit), and *Biospheres* (McGraw-Hill).

Robert Sternberg is IBM Professor of Psychology and Education, Yale University, and author of several books including *The Triarchic Mind: A New Theory of Human Intelligence* (Viking) and *The Triangle of Love* (Basic).

Francisco J. Varela is a biologist who holds the Foundation de France Chair in Cognitive Science and Epistemology at the Ecole Polytechnique in Paris. He is the author of *Autopoiesis and Cognition: The Realization of the Living* (with Humberto Maturana; D. Reidel) and *The Tree of Knowledge: The Biological Roots of Human Understanding* (with Humberto Maturana; New Science Library).

Hao Wang is professor of logic, Rockefeller University, and author of *Beyond Analytical Philosophy: Doing Justice to What We Know* (MIT) and *Reflections on Kurt Gödel* (MIT).

About the Editor

JOHN BROCKMAN, founder of The Reality Club and editor of *The Reality Club* series of books, is a writer and literary agent. He is the author of *By The Late John Brockman* (Macmillan, 1969), *37* (Holt Rinehart Winston, 1970), *Afterwords* (Anchor, 1973), and editor of *About Bateson* (Dutton, 1977).